The WWI French Aces Encyclopedia

Volume 1 | Achard to Boyau

David Méchin

This encyclopedia is dedicated to Christian-Jacques Ehrengardt (1949–2018), one of the greatest French aviation historians who showed me the way.

Acknowledgements

The author would like to thank his many friends and WWI aviation enthusiasts all around the world for his many exchanges of information and photographs: Damien Accoulon, Michel Bénichou, Frédéric Bruyelle, capitaine Sylvain Champonnois, Albin Denis, Robert Feuilloy, Didier Guénaff, Benoit Henriet, Philippe Guillermin, Lucien Morareau, Alexis Rocher, Claude Thollon-Pommerol (France), Colin Huston, Andy Kemp (England) and many gentlemen form Cross and Cockade, Michael Zeh (Germany), Paolo Varriale (Italia), Boris Ciglic (Serbia), Andrzej Olejko, Piotr Mrozowski (Poland), Ian Burns (Canada), John Guttman, William Nungesser, Greg VanWyngarden, Jim Wilberg (USA), Bernard de Broglio, and Colin Owers (Australia). Corrected edition.

For our aviation books in print please see our website at: **www.aeronautbooks.com**.
For questions or to help with photographs please contact Aeronaut Books at **jherris@me.com**

ISBN: 978-1-953201-30-0

Design and layout: Jack Herris
Cover design: Aaron Weaver
Digital photo editing: Jack Herris

Aeronaut Books
Books for Enthusiasts by Enthusiasts
www.aeronautbooks.com

Table of Contents

Introduction

*- Killing one guy while scaring the hell out of the whole regiment, that's the real job. And we did it, the job, with maybe three thousand guys. And we won it, this f***ing war! Us! The three thousand!*
- And the others?
- The others, they did it.

This dialogue comes from Bertrand Tavernier's masterful film adapted from Roger Vercel's 1934 novel of the same name, *Captain Conan*. This captain, played by Cesar-winning actor Philippe Torreton, asserts to his friend Lieutenant Norbert, played by Samuel Le Bihan, that it was men like those of his Corps Franc company who won the war: a very small minority of particularly motivated soldiers who won the battles by their decisive action, while the great mass of combatants merely held the trenches and, more rarely, went up blindly at the whistle, massing without really having the enemy at the end of the gun.

Whatever the relevance of these remarks about the infantrymen of the First World War, they are perfectly appropriate to describe the French fighter pilots, where most of the results were achieved by a very small minority, the "aces of aviation", an unofficial title given to any pilot or machine gunner who had obtained 5 certified aerial victories. 175 French pilots and gunners, plus 9 foreign pilots flying in French squadrons, officially obtained this status during the war with their air victories published in the Journal Officiel. Six other pilots who did not officially achieve this status, for various reasons, also deserve to be mentioned and will be included in the last volume. The encyclopedia you have in your hands will give you the most complete biographies possible of these 190 exceptional fighters, a very small phalanx of fighters whose military action was decisive in the conduct of the air war. Let's first try to put them in context to understand their importance.

1 - A brief history of French fighter aviation during the First World War

If the Wright brothers can claim the honor of being the first to have flown a heavier-than-air aircraft in controlled flight, France very quickly became the first nation in the world to welcome the exploits of the first air pioneers at the end of the first decade of the 20th century: Louis Blériot, the Caudron brothers, the Farman brothers, Léon Levavasseur, the Morane brothers, the Nieuport brothers, Gabriel Voisin... So many names of pioneers who will quickly pass from the artisanal stage to the industrial stage at the beginning of the 1910's, finding a number of engine manufacturers to equip their aircraft such as the Gnome company of the Seguin brothers, or the Renault company: France was at that time the first nation in the world for the production of automobile engines and would very logically become the first for airplane engines.

The French military was very quickly interested in the

Below: French top living aces in late 1918 - Nungesser - Madon - Fonck - Barny de Romanet in clear uniform - Heurtaux- 198 victories for those 5 pilots

Above: French front map in June 1915 with the first french fighter units.

possibilities of the new airplanes and benefited from the enthusiasm of the public, which offered them unlimited funds to equip themselves and increase their power via a national subscription scheme managed by the Ligue Aéronautique (Aeronautical League) headed by René Quinton. The first military aircraft were purchased in 1909, the same year that a military establishment was opened in Vincennes by Lieutenant-Colonel Jean Estienne, who was in charge of testing the possibilities of this new weapon.

If the first source of interest for the military was aerial reconnaissance and then adjusting artillery fire, Estienne and his collaborators quickly understood, during military maneuvers, that an aeroplane could very well intercept those of the opposing camp and prevent them from conducting reconnaissance.

Estienne then undertook aerial firing tests, led by Captain Charles Tricornot de Rose. Shooting was conducted from a two-seater Farman aircraft carrying a machine gunner. Conclusions quickly came that only a shot fired in the axis of the aircraft would allow a target to be hit with a minimum of accuracy. Estienne deduced that a weapon fixed on the engine cowling and firing forward was the best solution, but the problem of firing through the propeller remained. Captain Georges Bellanger, officer of the center, tested a Blériot XI equipped with a 37 mm revolver gun firing through the hub of the propeller. The aircraft disintegrated on the ground after the first shot... De Rose would carry out tests with Roland Garros at the center of Vincennes during the summer of 1912 on a Morane monoplane on which they fixed a machine gun on the engine cowling. They found that 10% of the bullets fired hit the propeller, and concluded that armoring the blades could be a solution. Unfortunately for their experiments, the center of Vincennes was dissolved in September 1912 for obscure political reasons and they both left for other activities. In parallel to their work, in April 1914, Raymond Saulnier, one of the partners of the Morane-Saulnier firm, registered a patent to synchronize firing a machine gun with a propeller. However, the patent could not be used operationally due to the poor quality of the ammunition of the time, which often caused delays in the ignition of the bullets.

When the war broke out on August 2, 1914, the French air force, perhaps the most operational of the belligerents, was used for reconnaissance missions and above all for artillery adjustment, a decisive advantage in the battle of the Marne, which enabled the French army to stop the progress of the German army in September 1914. After this battle, trench warfare began... And on October 5, 1914, a Voisin of the V 24 squadron, piloted by Sergeant Joseph Frantz and taking along as observer Private Louis Quenault, managed to shoot down a German two-seater Aviatik B that crashed in the region of

Above: Morane Parasol (Type L) 80 hp engine first improvised fighter plane early 1915.

Above: Morane Monocoque (Type N) with 80 hp engine summer 1915 with armored propeller blades.

Reims, in front of the eyes of General Franchey d'Esperey's staff.

This victory had such a great impact that General Joffre, in a note dated October 8, 1914, asked for the creation of one squadron per army (and two in Paris) to carry out aerial attacks against enemy aircraft. The first of these squadrons, MS 12, would be experimented within the 5th army in the region of Reims, at the initiative of the commander de Tricornot de Rose. Due to the lack of suitable aircraft for interception, it was equipped with the fastest two-seater aircraft of the time, the Morane Parasol, carrying an observer armed with a simple rifle behind the pilot. Five air victories were obtained by the squadron in April 1915, and fighter squadrons were becoming commonplace at that time. From west to east, there were MS 26 (8th Army, Dunkirk), MS 15 and MS 57 (10th Army, Brias), MS 23 (2nd Army, Petit Camon near Amiens), MS 3 (6th Army, Vauciennes), MS 12 (5th Army, Muizon near Reims), MS 38 (4th Army, Châlons-en-Champagne), MS 37 (3rd Army, Pierrefitte), MS 31 (1st Army, Toul), MS 48 (Detachment of the Armies of Lorraine, Lunéville) and MS 49 (Army of the Vosges, Fontaine, near Belfort and the Swiss border).

While these squadrons were being formed, the first fighter aircraft also arrived. The famous aviator Roland Garros, mobilized into the MS 23 squadron which fought in Lorraine

Above: Nieuport 10 first standard single-seater fighter with 80 hp engine during mid 1915.

Above: Nieuport 11 standard fighter with 80 hp engine during the battle of Verdun on February 1916.

in August 1914, was asked at the end of the year to resume his experiments of shooting through the propeller started two years earlier at the Vincennes Center. Failing to find a synchronization device for shooting through the propeller disc that was really effective, he at least installed a propeller equipped with armor on the blades of this one, which deflected bullets that would strike it. He concluded that "*The light, powerful, fast single-seater is currently the only aircraft capable of catching up quickly, in distance and altitude, with enemy aircraft, and once caught up, of imposing on them, through superior maneuverability, the maneuvers most conducive to attack. It was a question of knowing if one could practically arm a single-seater with a propeller at the front, the only device giving the necessary speed and efficiency.*" Finally, in February 1915, he flew a Morane Parasol as a single-seat aircraft to test his invention at the front, at the MS 26 squadron in Saint-Pol sur Mer, near Dunkirk. He won three aerial victories aboard his

Above: Nieuport 16 - a more powerfull version of Nieuport 11 with a 110 hp engine - appearing on march 1916.

Morane Parasol in April 1915, before being forced by ground fire to land in enemy lines where he was captured.

The General Staff ordered ten Morane "Garros" aircraft from the Morane company, as well as an equivalent number of Morane Monocoque aircraft, with which the famous pre-war aviator Eugène Gilbert had some success in June 1915. But neither of these two aircraft would be produced in large numbers, because in May 1915 a small biplane, the two-seater Nieuport X, entered service. It was quickly converted into a single-seater fighter with a fixed machine gun mounted above the upper wing, firing outside the propeller's arc. Its smaller, lighter, and improved version, designed specifically for hunting, the Nieuport XI B (known as Bébé) entered service at the end of 1915 and became the standard aircraft of the French fighter pilots.

The latter was to have its moment of glory at the time of the Battle of Verdun, which began on February 21, 1916, when most of the French fighter resources were to be deployed in this sector under the sole command of Major de Tricornot de Rose. These fighters, by disrupting the work of the enemy artillery adjustment planes and by securing the work of the French planes, brought decisive help to the French artillery, which was then outnumbered by its German counterpart, and allowed the French army to win the defensive battle.

The French fighter squadrons fought again at Cachy, near Amiens, under the command of Major Antonin Brocard during the Battle of the Somme, initiated to relieve the pressure on Verdun. On their new Nieuport 16 and 17 fighters equipped with 110 horsepower engines, and, at least for the Nieuport 17, with a synchronized machine gun firing through the propeller arc, it was a victorious period in the face of a German air force which could only oppose them with the now obsolete Fokker Eindecker fighter. The latter reformed and created its first fighter squadrons, the Jastas, in August 1916 and equipped them with biplane fighters, of mediocre quality from Fokker and Halberstadt, then of good performance for the Albatros D.II and then D.III with their powerful 160 horsepower engines, which made them superior to the Nieuports.

Unfortunately for the Germans, this was also the period when the excellent SPAD VII, with its 150 horsepower Hispano-Suiza engine, entered service in the French air force. Constantly improved during the war, with the arrival of a 180-horsepower over-compressed version in early 1917, followed by the even more powerful SPAD XIII with two machine guns, the SPADs gave the French air force technical superiority oveghout the rest of the war – only the last Fokker D.VIIs with over-compressed BMW engines were able to counter them effectively, but too few of them entered service at the end of the war to be of decisive use.

In addition to this technical superiority, the French air force was able to rely on its increasingly grouped use of aircraft as the war wore on, always one step ahead of the German air force. The first French fighter squadrons were created in March 1915, more than a year before the Jastas of the German air force, which until then had dispersed its fighters in two-seater units, even if the Fokkers coming from various squadrons used

Above: Nieuport 23 - a late version of the Nieuport 17 durig year 1917. - Photo Fernand Cuville

to work together in Kampfeinsitzer Kommandos (KEKs). In October 1916, the first four French Combat Groups (GC), permanent groupings of four squadrons that fought together and shared the same airfields, were created. The GCs 11, 12 (Cigognes), 13, and 14 were created eight months before the first Jagdgeschwader, its German equivalent, which was only created in June 1917. While only five Jagdgeschwaders were created by the Germans by the end of the war, the French created a total of 13 Combat Groups, six of which, the GC 15, 18, and 19 (1st Wing) then 13, 17, and 20 (2nd Wing), were to be grouped together at the beginning of 1918 within the Air Division, an armada under the direct orders of the Grand Quartier Général that was immediately sent to all the sensitive points of the front to seize air superiority from the enemy. This mass of fighters was to prove to be a decisive asset to face the last German offensives of the spring of 1918, and then to powerfully support the Allied counter-offensives organized during the summer, which were to take back the lost ground and push the German army back to Belgium until the armistice of November 11th 1918.

2 - The weight of the French air force among the Entente's air forces

When the fighting stopped, the French air force was the most important of all the belligerent nations air services, after an exponential increase in power since 1914, especially in the fighter sector, with no less than 72 squadrons, each theoretically consisting of 18 SPAD fighters, i.e. **1,296 single-seat fighters** in line. The German air force, on the other hand, only fielded 86 Jastas of 14 fighters (in theory, usually less in fact), i.e. only **1,204 single-seat fighters**, which had to fight, in addition to the French, the **1,080 British fighters** (45 fighter squadrons with 24 aircraft), the **288 American fighters** (16 Pursuit squadrons each with 18 aircraft) and the **54 Belgian fighters** (3 fighter squadrons with 18 aircraft), i.e. a theoretical total of **2,718 Allied fighters against 1,204 Germans on the Western Front alone, i.e. a superiority of 2. 25 to 1.**

The French represented almost half (47.6%) of the Allied fighters on the line at the armistice. Not surprisingly, they also won almost half of the air victories on the Western Front.

The comparison of the results obtained by each air nation is a delicate exercise, due to the different Confirmation system between each nation. The British had a very lax system, based on the pilot's declarations, a system that served above all to boost the morale of their air force, which had been decimated several times by enemy aircraft. The French system was much stricter, based on the proof of the destruction of the enemy plane, either by finding its wreckage, or by collecting testimonies of its destruction from troops outside the ones claiming victory. It would be necessary to compare these claims with enemy losses; however, we do not have statistics on aircraft losses in the German and French air forces.

The only statistic common to the three major belligerents is that of pilots lost to the enemy, killed or taken prisoner. These statistics have been published in various books during the 2000s: *Casualties of the German Air Service 1914–1920*, by Norman Franks, Franck Bailey and Rick Duiven (Grub Street, 2008), *The French Air Service War Chronology 1914–1918* by Franck Bailey and Christophe Cony (Grub Street, 2002), and *The Sky their Battlefield: Air Fighting and Air Casualties of the Great War – British, Commonwealth and United Air Service 1912 to 1919*, by Trevor Henshaw (Fetubi books, 2014 for the 2nd edition).

Above: SPAD VII the fastest fighter plane with 150 hp then 180 hp engine after September 1916.

If we carry out the tedious exercise of counting pilot losses, we can see that the German air force lost 2,351 pilots (killed or captured) on the Western Front from 1914 to 1918. The French army lost 1,285 during the same period, and the British army (RFC, RNAS, and RAF) lost 3,238. The question remains as to how many German pilots fell to the British forces and how many to the French forces?

This question can be answered because we know the overall location where each German pilot fell or was captured. This makes it possible to establish the geographical distribution and, knowing the separation of the French and British sectors (the Somme River for most of the war), to determine approximately the proportion of the losses to be imputed to the French or British forces. However, an important correction must be made for the Battle of Flanders, which took place from July to November 1917, and which was fought in the British sector, but with important fighter reinforcements brought in by the French. If there were 18 British Squadrons of 18 fighters each (324 aircraft on the line), there were also 13 French fighter squadrons of 15 fighters each (195 aircraft on the line) for the months of July and August; reinforcements dropped to 9 squadrons (135 fighters) for the months of September, October, and November 1917. From this observation, 38% of German losses on the British front in July and August, and 29% of those from September to November were attributed to the French air force in our statistics. From these observations, we arrive at the following casualty table for the major belligerents on the Western Front:

These figures are obviously only approximations. Some of the pilots lost were not because of the fighters, but because of fire from the ground, or even because of mechanical problems. But we can assume that the same is true for all nations, and in the same proportions.

From this analysis, it appears that the French air force was

Aviation Casualties on the Western Front

	French Front			British Front		
Year	**German**	**French**	**U.S.**	**German**	**British**	**U.S.**
1914	4	11		-00	15	
1915	31	58		19	80	
1916	136	231		81	300	
1917	379	292		366	1,164	
1918	706	693	200	629	1,679	27
Total	**1,256**	**1,285**	**200**	**1,095**	**3,238**	**27**

Right: Casualty data show that French to German losses were about 1:1 wheras British to German losses were about 3:1.

Above: SPAD XIII enhanced version of SPAD VII, with a 240 hp engine and 2 machine guns, seen here in escadrille SPA 84 in 1918.

responsible for a little more than half (53%) of the German losses on the Western Front, even if a small part of the losses were inflicted by American pilots – a sum estimated at about sixty German pilots (3%), taking into account the late engagement and the small number of American units.

Another lesson is that the French and German air forces inflicted roughly equivalent losses on each other (1 German for 1.02 French), while the British suffered heavily from the German air force, which inflicted almost three times as many losses (1 German against 2.96 British) in terms of pilots killed or taken prisoner.

British aviation suffered for a long time from obsolete equipment and above all from very neglected training of pilots, which led to the hecatomb of the Battle of Flanders in 1917, where an infernal spiral started: to compensate for the losses, the head of the Royal Flying Corps, Sir Hugh Trenchard, took the decision to cut back on the training time of the pilots...

3 - French pilots among the best equipped and trained

French aviation never had to suffer from these difficulties. As indicated in the history above, from the beginning of the war until the armistice, it had at its disposal, at least for fighter aircraft, high-performance equipment such as the Nieuport and SPAD fighters, all of which were produced in large quantities and equipped the air forces of all the Entente nations.

If the French pilots had the best equipment, they also benefited from the best training. The system of French flying schools was entrusted on September 19, 1915 to the congressman Adolphe Girod (1872–1933), a reserve commander who was promoted to the rank of colonel at the end of the war. He organized the training curriculum for pilots and was helped by the fact that many civilian schools existed, created before the war by the numerous French aircraft manufacturers. These schools were militarized and doubled in size with new schools created during the war.

Girod, whose clumsiness in piloting was proverbial, was not satisfied with managing an already existing system, but made sure that this new one could train more pilots than necessary compared to the needs in the squadrons, which trained many pilots, who already had a minimum of 40 hours of flight in their training course. This means that many pilots, who already had a minimum of 40 hours of flight time in their training program, left the program without having a squadron assignment and had to wait for a place to become available, while continuing to train at the General Aviation Reserve (RGA) at Le Bourget or at the Groupement des Divisions d'Entrainement (GDE) located at Le Plessis-Belleville, near Paris. It was not uncommon, when comparing the few flight logs that have come down to us, for some fighter aces to have more than a hundred hours of flight time under their belt when they finally arrived in their unit, a decisive advantage over the enemy.

At no time during the war was the French Air Force short of pilots, even during the worst moments of the 1918 offensives, when losses were the highest. The worst month of the war for the French air force was June 1918, when 123 pilots were killed in action. That same month, 526 pilots graduated from French flying schools... There was never the slightest difficulty, throughout the war, to provide the pilots necessary

Right: Lt-Colonel Girod, the head of french aviation schools during WWI.

for the growth of squadrons.

16,898 military pilots' licenses were awarded during the war in French flying schools, some of them to foreign pilots. They are added to the 492 pilots who received their license before the war, between February 7, 1911 and July 31, 1914. The number of fighter pilots among all these pilots can only be estimated – given the proportion of fighter pilots at the time of the armistice (42%), it is probably around 7,300.

All these pilots won a total of 2,817.93 certified victories during the war, a figure obtained by adding the 2,694.50 victories recorded by historian Franck Bailey on the French front to the victories won by French pilots on other fronts (75.5 on the Eastern front at Salonika as well as during the Serbian and Dardanelles campaigns, 20.93 on the Italian front, 14 on the Romanian front, and 13 on the Russian front).

The 184 or so official aces of the French air force, who are in fact only 177 pilots if we exclude the seven machine gunners, won exactly 1,322.36 certified victories, i.e. nearly 47% of the total, while they represent a little less than 2.5% of the fighter pilots! They therefore represent a military reality even before being propaganda creations.

Above: Croix de guerre with 3 palms 1 vermeil 1 silver and 1 bronze star.

4 - The aces: a small group of over-motivated pilots

From the study of the files of the fighter aces, we can see that all of them were born between 1885 and 1898, which gives them an age range between 16 and 29 years old in 1914. Their average age was 23 years old, only 9 of them were just over 30 years old when the war broke out. Like all the men of their generation, they were educated by the public schools in the spirit of revenge against Germany, with their eyes turned towards the blue line of the Vosges, but their patriotism was going to manifest itself in a much more important way.

If we take only the 175 aces of French nationality, when the posters of general mobilization were posted in all the communes of France, 51 (30%) were active military personnel who were by definition already in the armed forces, in various arms. 17 others were in the process of performing their national service and were also already wearing the uniform. The remaining 107 would be mobilized when war was declared, but 42 of them (39%) would not only wait for their mobilization order but also voluntarily enlist in the army. This is a considerable proportion, given that for the entire armed forces, which numbered nearly 3.5 million mobilized soldiers, only 100,000 voluntary enlistments were recorded during the war – 2.8% of the workforce.

This particularly strong patriotism can be explained by the social origin of the people concerned, where there is a very large overrepresentation of the upper classes where nationalism, even militarism, is a widely shared value. If we look at the father's profession recorded on the birth certificates, we see that almost half of them (85) come from the upper classes, with the sire working as a career officer, doctor, academic, industrialist, or rentier. About a quarter (45) came from what could be called the middle class with professions such as small businessman, civil servant, or journalist, while only a quarter (45) came from the working classes (farmers, workers, employees, or craftsmen). It should be noted that in 1914, farmers alone represented about 55% of the active population, but barely 9% (15) of the original population of the aviation aces.

Whether their incorporation into the ranks of the army was voluntary or involuntary, all of them, without exception, volunteered to join military aviation. Very few of them (18) were already licensed military or civilian pilots before the war. Most of them learned to fly during the war, which at the time was done fairly quickly, in about six months. We note three concrete elements in their files, which are sometimes cumulative, and which can explain their motivation.

First of all, like all men of their generation, they were fascinated during their adolescence, that is to say during the first decade of the 20th century, by the development of mechanical sports. The press reported the first automobile races, as well as the flights of the first daisy cutters, whose flight over the smallest village by an airplane was a local event. A third of them would transform this passion into a profession by becoming, at the dawn of the 1910s, engineers, mechanics, car drivers (which at the time was more or less the same thing...), or employees of the aeronautical industry.

The second reason, which was less obvious, was the desire to escape the trenches. Many soldiers made this request in 1918 in the hope of obtaining a six-month respite from the front and with the additional motivation of benefiting from much more comfortable living conditions. The combat value of these pilots was, according to several testimonies, quite relative... But the aviation aces, if they did not disdain the advantage of the comfort given to the pilots, had above all the will to continue to fight in a way considered more flamboyant. 55 of them (31%) started the war in the cavalry, the most prestigious arm, but which quickly became obsolete because of the trench warfare that transformed it into second line infantry. For them, it was as much a question of leaving the mud of the trenches

Above: Officer and knight cross of légion d'honneur.

and the uninteresting missions as it was of changing their mount.

Finally, it should be noted that 20% of the 175 French aces were "éclopés", soldiers judged unfit for the infantry because of wounds received at the front or the weakness of their constitution at the time of their incorporation. These men could have spent the rest of the war at home or on secondary tasks far from the front line, but made the deliberate choice to fight and to join the air force, where the physical criteria were much more flexible.

5 - Medals, glory, and privileges

In a squadron, fighter aces generally formed a very small minority of pilots, as ace Louis Risacher pointed out: "*In combat, you have to be extremely motivated to go and kill a man. So, if in a unit, you have 10 percent of the pilots who want to fight, that's already good. A lot of pilots were just there to have a nice uniform. In combat, you have to go for it. As soon as you see the enemy, you have to dive on them, and improvise to get close. Then, not everyone being like Fonck, shoot at close range.* "With the notable exception of the legendary escadrille SPA 3, an elite unit with many aces, in almost all the other squadrons, most of the collective results of the unit were the result of one, two, or even three pilots, knowing that the regulatory strength was set at 18 at the end of the war. The

Above: Médaille militaire.

ace of aces, René Fonck, personally won 72.5 of the 103.5 victories of escadrille SPA 103 (split up to take into account the victories shared with other units). Charles Nungesser alone won nearly a third of the 90.84 victories of SPA 65, and the proportion rises to more than half if we add the victories of the seven other aces of the unit, Robert de Bonnefoy, Jean Sauvage, Marcel Henriot, Eugène Camplan, Georges Lienhard,

UN JEUNE HÉROS DE L'AVIATION

LE VAINQUEUR DES FOKKERS

Pilote et mitrailleur, Guynemer accomplit prouesses sur prouesses

Chez la grand'mère du héros du «Vieux Charles»

Le Petit Parisien a relaté, le premier, les exploits du « Vieux Charles » et du héros qui le monte. Mais il ne nous a été permis qu'hier de révéler son nom : Guynemer. Nous l'avons dit : celui qui, du premier

UN HÉROS DE L'AIR

LES EXPLOITS DU SERGENT GUYNEMER

Ce pilote de vingt et un ans, titulaire de la Légion d'honneur, de la Médaille militaire et de la Croix de guerre avec quatre palmes, vient d'abattre son cinquième avion

Chez la grand'mère du pilote Guynemer

LE SERGENT PILOTE GUYNEMER

UN AVION-CANON ABAT UN DRACHEN

553e JOUR DE LA GUERRE

Above: Press coverage of Guynemer becoming an ace.

Jacques Gérard, and Lucien Cayol, knowing that 124 pilots succeeded one another in the squadron during the war. Less well known, the two aces of SPA 75, William Hérisson and Antoine Laplasse, won 13.41 of the 31.41 victories of the squadron, i.e. 43% of the total.

A confirmed victory is defined as any enemy aircraft, captive balloon, or airship destroyed or captured and recognized as such by the military authorities through a Confirmation decision. The victory, contrary to what is sometimes written, did not necessarily have to fall in the French lines (most of them did not) but could fall in the enemy lines, provided that sufficient evidence of the enemy's downfall was collected.

The consequence of this confirmation decision, which is taken at the level of each of the French armies, is in the vast majority of cases the rewarding of the victorious pilot or machine gunner by a citation in the Order of the Army (OA) which is printed in the columns of the Journal Officiel de la République Française (JORF), whose number and page are specified in the biographies of this encyclopedia. There are, however, cases where the victory is rewarded only by a citation of lower rank, in the Order of the Army Corps (OCA) which do not appear in the Official Journal and must be noted only in the individual files of the combatants.

Every citation gives the right to a decoration, the first of which is the Croix de Guerre. Created by the law of April 2, 1915, it has a bronze palm on its ribbon for any citation awarded to the order of the army, then a vermeil star for a citation to the order of the army corps, a silver star for a citation to the order of the division, and a bronze star for a citation to the order of the regiment. The greatest aces, who received many citations, therefore had war crosses with a particularly rich ribbon. The most important one is that of the ace of aces René Fonck with 25 palms and a bronze star. Nungesser's cross, which had only 19 citations in the army order and 2 in the army corps order, showed a ribbon with 28 palms and 2 vermeil stars... He had awarded himself as many palms as foreign decorations, which did not, however, confer any palms on his war cross. No one allowed him to make the remark and it seems in fact that the length of the war cross was also a reflection of the ego of those who wore it... Fonck, who, after the war, was promoted to Commander and then Grand Officer of the Legion of Honor, added an additional palm to himself on each occasion, bringing his total to 27, even though these promotions did not entitle him to them. He is also the only one to use the silver palms, which are worth 5 bronze palms.

5 heures du matin — ÉDITION DE PARIS — 5 heures du matin

Le Numéro 10 Centimes

TÉLÉPHONE
Administr. Gut. 01.67-01.74-01.75
Rédaction. Gut. 01.76-01.77-01.78

Le Petit Journal

SAMEDI * 11 * MAI
SAINT MAMERT
56me Année — 1918 — N° 20,224
61, rue Lafayette, Paris (9e)

LA LEÇON des travaillistes américains

Loin d'exagérer l'esprit de guerre dont sont animés les travaillistes américains qui vont nous quitter aujourd'hui, la presse française, en y applaudissant, n'en a pas rendu toute l'intensité ni souligné toute la vigueur. Je viens d'avoir le « privilège », comme on dit là-bas, d'accompagner ces vrais démocrates dans leur visite aux régions du front, et leurs paroles que, de leur consentement, je vais fidèlement rapporter me dispenseraient presque de tout commentaire.

John P. Frey, d'abord, qui improvisa sur le bateau l'étude du français et, seul, le parle à sa rude et expéditive façon, broyant la grammaire, défonçant la syntaxe, l'idée volontaire surmontant le but comme un char d'assaut. Au général commandant une de nos armées : « *Notre résolution*, dit-il, *de ne pas nous rencontrer avec le Boche* (sic) *s'est faite plus inébranlable encore depuis notre visite à Reims.* » Et dans cette ville même, lorsque les officiers de service nous introduisirent au chevet de la grande morte de pierre, cette explosion d'une colère physique : « *J'ai un fils soldat en France ; si les Allemands me le tuent, je ne pourrais me défendre de l'instinct de vengeance qui me ferait chercher une baïonnette pour en transpercer le meurtrier ; eh bien ! c'est cette rage que j'éprouve ici devant le meurtre de cette cathédrale.* » Puis, à Verdun, au colonel qui commande la place, il jette un joli trait d'imagination, fixant la hantise de la pensée de guerre ; sur les ruines de la vieille *Firmitas*, la citadelle du temps des Romains qui transmit le mot d'ordre à l'avenir, Frey aperçoit un cartouche des rois de France au nom de *Louis*, dont les obus ont arraché ou mutilé

SUCCÈS FRANÇAIS A GRIVESNES

Le parc repris -- 258 prisonniers

COMMUNIQUÉS FRANÇAIS — *10 Mai, 14 heures.*

Après une intense et brève préparation d'artillerie, nos troupes, dans l'après-midi d'hier, se sont emparées du parc de Grivesnes dont une importante partie restait occupée par les Allemands.

Au cours de cette opération, nous avons fait 258 prisonniers dont quatre officiers, et ramené un nombreux matériel.

Malgré les vives réactions de l'artillerie ennemie et les reconnaissances qui tentaient d'aborder notre nouvelle ligne, nos fantassins se sont maintenus sur les positions conquises et les ont organisées.

Sur la rive droite de l'Ailette ; en Champagne (région de Massiges et nord de Reims) ; en Lorraine (bois d'Ailly), nous avons exécuté avec succès plusieurs opérations de détail ou repoussé des coups de main ennemis au cours desquels nous avons fait 36 prisonniers dont un officier.

10 Mai, 23 heures.

Rien à signaler en dehors d'une lutte d'artillerie assez violente dans la région de Grivesnes et entre Lassigny et Noyon.

COMMUNIQUÉS BRITANNIQUES — *10 Mai, après-midi.*

Le faible élément de tranchées de première ligne, situé au nord-ouest d'Albert, où l'ennemi avait pénétré par suite de son attaque, hier matin, a été repris par notre contre-attaque pendant la soirée. Nous avons fait quelques prisonniers.

La nuit dernière, l'artillerie ennemie s'est montrée active dans les vallées de la Somme et de l'Ancre et en différents points du front de la Lys.

10 Mai, soir.

Des combats locaux ont eu lieu dans le bois d'Aveluy au nord d'Albert.

En dehors de l'activité de l'artillerie ennemie sur différents points, particulièrement dans les secteurs de Vimy, de Robecq et au sud de Dickebusch, il n'y a rien d'autre à signaler.

DE VIOLENTES ATTAQUES avant la grande bataille

Des actions locales d'où résultent diverses fluctuations se développent en grande majorité à l'avantage des armées anglo-françaises. Le front en est peu affecté dans son ensemble ; les déplacements en profon-

FONCK ABAT six avions boches en une journée

dont trois en moins de six minutes

(Officiel français.) — *10 Mai, 14 heures.*

Dans la journée d'hier, le lieutenant Fonck, au cours de deux patrouilles, a abattu six avions bi-places allemands : les deux premiers en dix secondes, le troisième cinq minutes après, les trois derniers au cours de sa deuxième patrouille.

(Cl. H. Manuel.)
FONCK, l'" as des as "

Les 6 avions de Fonck

Le lieutenant Fonck a accompli le 9 mai

QUAND RENTRERONT les prisonniers de guerre

Les mesures de représailles suspendues. Plus de camps de concentration pour civils. Plus de déportations dans les pays occupés.

Nous avons donné d'une manière très complète les bases essentielles des accords conclus à Berne au sujet de l'échange des prisonniers de guerre. La communication officielle qu'on trouvera plus loin précise les conditions de cet échange qui doit s'étendre aussi aux civils internés ou retenus dans les deux pays.

Il résulte de cette communication que tous les prisonniers de guerre internés en Suisse avant le 20 avril comptant 18 mois de captivité seront rapatriés à partir du 15 mai.

Les prisonniers de guerre restés en Allemagne et qui remplissent les conditions pour bénéficier des dispositions de l'accord ne commenceront à être rapatriés que dans le courant de juin.

Quant aux civils ils seront tous rapatriés, et ceux résidant dans les régions envahies pourront même rentrer en France. C'est la suppression complète des camps de concentration.

D'autre part, il convient de remarquer que toutes les mesures de représailles ont été de part et d'autre immédiatement suspendues.

ACCORD sur les prisonniers de guerre

Les accords récemment conclus à Berne entre les délégués du gouvernement français et ceux du gouvernement allemand viennent d'être définitivement ratifiés par les deux gouvernements. Ils entreront en vigueur dès le 15 mai.

L'échange des prisonniers de guerre français et allemands a pour base l'ancienneté de capture. En principe, tous les

ACCORD concernant les civils

L'accord concernant les civils comprend deux parties :

La première vise la libération de tous les civils actuellement internés en France et en Allemagne. C'est la suppression de tous les camps d'internement civils. Les Français qui s'y trouvent actuellement seront renvoyés en principe en France libre si leur dernier domicile était en Allemagne ou en France libre ; dans les départements envahis si leur résidence se trouve dans les régions encore occupées par les troupes allemandes. Ces derniers peuvent toutefois demander à être envoyés en France libre et il sera donné satisfaction à leur demande si des nécessités militaires ne s'y opposent pas. D'autre part, les Français qui ont été, à une époque quelconque, internés et qui ont par la suite obtenu l'autorisation de résider librement en Allemagne, bénéficieront de l'accord s'ils en expriment le désir dans les mêmes conditions que s'ils étaient encore internés. Il en est de même réciproquement pour les civils qui se réclament de la nationalité allemande et qui sont ou ont été placés en France dans des dépôts d'internement. Ils peuvent obtenir, selon le cas, l'autorisation de rentrer dans leur foyer ou de se rendre en Allemagne. Les civils mobilisables qui rentreront, par application de l'accord, en France ou en Allemagne ne peuvent être employés à un service militaire ni au front ni dans la zone des armées, ni sur les territoires alliés.

Le traitement des populations envahies

La seconde partie de l'accord est relative au traitement des populations des régions occupées. Laissant de côté toutes les questions de principe, elle précise, en fait, la Convention de La Haye à laquelle elle se

Above: *Le Petit Journal* headline May 11 1918 - Fonck shoots down six enemy planes

Two other more traditional decorations were awarded, the first being the Legion of Honor, created in 1802 by Napoleon Bonaparte. The knight's cross, the first rank of the order, is generally awarded to any officer who distinguishes himself in combat but can also be awarded to a non-commissioned officer. The higher rank, that of officer (with a rosette on the red ribbon), was given to only 12 of the 175 French aces during the war – several received the higher ranks of the order, but well after the war. The second traditional decoration is the military medal, created in 1852 by Napoleon III, is intended to reward soldiers and non-commissioned officers who have distinguished themselves in combat. Some of the French aces, who were already officers at the time of their exploits, did not receive this decoration. It should be noted that the citation, published in the Journal Officiel, granting the military medal or the legion of honor, counted as a citation in the order of the army and thus gave the right to an additional palm on the war cross.

In addition to the numerous decorations, the Grand Quartier Général (GQG), and most probably Major Joseph-Edouard Barès, head of the GQG's aeronautical department, decided at the beginning of 1916 to honor the best fighter pilots in a very special way: any pilot who had obtained five certified aerial victories in aerial combat would henceforth have his name cited in the daily communiqué of the armed forces, published at the beginning of the Journal Officiel. The first to receive this honor was the young Sergeant Georges Guynemer, on February 6, 1916, who was quoted in the following text: *"This morning, at about 11:30, Sergeant pilot Guynemer fought an enemy aircraft in the Friesland region and shot it down, in flames, between Assevillers and Herbécourt. This was the fifth enemy aircraft shot down by Sergeant Guynemer*". On the following February 27, it was Jean Navarre's turn to be distinguished in this way, followed by Charles Nungesser on April 21, followed by 70 others. Only 73 of the 184 aces actually received the honor: at the beginning of the summer of 1917, the GQG raised the threshold for inclusion to 10 air victories. 106 aces having obtained between 5 and 9 victories after this period would therefore not be distinguished in this way, as well as five others having reached this threshold a few weeks before the cut-off date of the armistice when the communiqué ceases to be published.

The fact of seeing one's name mentioned in the communiqué was enough to ensure its beneficiary an immense celebrity, multiplied by the action of the press. Subjected to military censorship, the press had only this one communiqué as a source of information for writing articles. The communiqué was generally not very verbose and provided information on artillery exchanges on this or that part of the front. Journalists then enthusiastically took hold of the airmen who were thus presented to them and who were practically the only combatants to benefit from this honor. Articles were devoted to them on the front page of the daily newspapers, sometimes with their photo, which was a privilege usually reserved for politicians or the most prominent artists. Knowing that the press was the only media available to the public at the time, one can imagine the immense celebrity of those concerned, whose relatives were sometimes harassed by journalists. Guynemer's grandmother was interviewed the day after her name was

1re Année. — N° 1 Le numéro : 50 centimes Jeudi 16 Novembre 1916.

La Guerre Aérienne Illustrée

RÉDACTEUR EN CHEF : JACQUES MORTANE

REVUE HEBDOMADAIRE
ADMINISTRATION : 30, Rue de Provence, PARIS

LA GUERRE AÉRIENNE AU JOUR LE JOUR

(Cliché Capitaine N. G...)

TROIS AS DANS UNE CARLINGUE

Ces trois champions de la chasse, habitués aux appareils fringants, souples et rapides, semblent un peu dépaysés dans cet avion-canon imposant, puissant et majestueux. Ce sont, de gauche à droite, les sous-lieutenants Guynemer et Heurteaux et le lieutenant Deullin, de l'escadrille des . A eux trois, ils avaient officiellement à leur actif, le 5 novembre, 36 appareils ennemis : Guynemer, notre recordman, 18 ; Heurteaux, 11 ; Deullin, 7.

Above: *La guerre aérienne illustrée* - the french aviation newsmagazine

mentioned in the press release, and she obviously told the journalists how proud she was of her grandson.

It was the journalists in the press who popularized the term "flying ace", which did not appear in any military document. This term comes from a popular expression referring to the greatest athletes. It was first used in an article by Georges Prade (1875–1920), a sports journalist and former editor of the newspaper *"Les Sports"* from 1904 to 1909. On April 13, 1916, in the daily newspaper "*Le Journal*", he wrote an article about the pilot Jean Navarre, then the best fighter pilot in French aviation. His article simply transcribed the comments heard in the world of pilots about him: "This one is the ace".

Let's mention in particular a weekly publication, "*La guerre aérienne illustrée*" (The illustrated air war), which appeared from November 1916 on the initiative of the sports journalist Jacques Mortane (Jacques Romanet for the civil status). The pages contain many articles on the exploits of the aviators, especially those of the fighter aces, who also benefit from a detachable poster portrait in the center of the magazine. For the young male and even female public, this publication was like baseball cards to collect, as Louis Demazières, an aeronautical engineer during the interwar period, who was a teenager during the Great War, testifies: "La Guerre Aérienne illustrée *arrived, with Jacques Mortane as editor-in-chief. As a student in high school, I received this magazine which I filed carefully. The first page of the first issue was for Guynemer. Guynemer, Nungesser, Heurtaux, Madon, Coiffard, all these great names I remembered! At that time we were keeping the daily list of winners, following the army's communiqué which appeared every day and reported the prowess of the aviators as soon as they had passed their 5th or later their 10th victory.* "

This fame was not without consequences on the daily life of those concerned, at least those who had passed the communiqué, who became the rock stars of the time. The ace Alfred Heurtaux says nothing less in the oral testimony that he left to the historical service of the defense: "*The number of invitations that we were able to receive... It was like the fashionable singers. We put the letters in a pile on the squadron table. Each of us knew which letter or letters they were interested in. And then the others, we would open them together. You can't imagine what we got, and what we received. In the end we were annoyed... In the evenings, when I went out with Guynemer, we found jewels in the pockets of our overcoats, with addresses. What's the point? I know that there are some who took advantage of it... Sometimes we had letters from children. So we would send them the gifts we received, exchange of goodwill.* "

The greatest aces were quickly caught up in their fame when they went out on the town on leave. While the SPA 3 squadron was stationed near Nancy in February and March 1917, its pilots were celebrated by the local authorities who organized a banquet in their honor and where aces such as Guynemer, Heurtaux, Dorme, and Deullin were chased by autograph hunters. The most media-savvy of the aces, Georges Guynemer, complained about this celebrity, which prevented him from having a quiet dinner in a restaurant where men wanted to talk to him to congratulate him and women wanted to kiss him. Fonck, while going to a Parisian cinema, was recognized by the public which rose to applaud him and rushed to congratulate him to the point that he had to make his way towards the exit surrounded by swarms of admirers... If the less known aces can undoubtedly move quietly in town, all those passed to the communiqué receive an abundant mail of admirers and also of admirers who send them perfumed letters. The journalist Jacques Mortane complained besides in one of the editorials of his review of the too numerous requests of disclosure of addresses of aces which he received. The ace of aces René Fonck was even the victim of an admirer to whom he had the misfortune of answering to see himself already married by the mother of the interested party who contracted debts in his name: the affair ended in the council of war where the ace was cleared but confessed to his comrades that he had almost been shot...

There were many gifts among the admirers' mail; the ace Charles Lefèvre would even receive one after the war at the end of 1919 when, coming back from a mission in Japan, he learned that it was written in the will of an old lady who wished to thank the pilots having shot down the Zeppelin L 49 on October 20, 1917 and left them a sum equivalent to 1,650 of our current dollars... Even more than private individuals, aircraft and engine manufacturers were generous with the aces, who were for them powerful relays of opinion with the military authorities deciding on arms orders. At the beginning of the war, the Michelin brothers in particular offered a million francs to the State, asking that it be distributed among the pilots and observers who distinguished themselves in combat; one understands from reading the readers' letters of the Mortane magazine that it was used to reward any aviator who had a certified victory. German propaganda got hold of the affair and the *Frankfurter Zeitung* did not hesitate to present the ace of aces Guynemer as a venal being receiving 1500 Francs (4,700 current US Dollars) for each aircraft allegedly shot down. This newspaper was underestimating the truth if we are to believe a letter from the Michelin brothers to the President of the Republic, dated October 6, 1914, in which they offered to pay 5,000 francs ($15,600) to any aviator who shoots down an enemy aircraft certified by the Ministry of War! Another Mortane editorial dated July 26, 1917, states that "For some time, bonuses have been forbidden" which does not seem to have been rigorously applied because a circular from the under-secretary of state Jacques-Louis Dumesnil was issued on February 20, 1918, forbidding pilots to accept any bonus from a manufacturer. The best illustration of this bonus system that made the fortune of the aces was the fact that Charles Nungesser after the armistice lived in an apartment on the Champs Elysées. Without being an expert on the real estate market of the time, one can easily assume that he could not have amassed a fortune during the conflict from his second lieutenant's pay alone, which amounted to 720 F per month ($2,250), including flight allowances. Abroad, the Italian government gave a bonus of 1,000 lire ($3,135) for each aerial victory, which the only French ace on this front, André Lévy, received.

Benefiting from an unequalled comfort of life compared to the poilu wading in his trench and from a well-stocked wallet and great prestige, the aviators during the leaves which are generously granted to them are the kings of the Parisian nights, and especially the aces of aviation. Airmen were forgiven many misdemeanors, especially Jean Navarre, the ace of the aces during the battle of Verdun, who liked to drown his stress in alcohol and pick fights. Depressed by the death of his twin brother, he committed the irreparable by driving his car into police officers in the streets of Paris. Arrested, he was judged irresponsible for his actions by the military court, which would certainly not have been so lenient for a simple soldier. The example of another ace, Charles Quette, is just as revealing: coming from the working-class districts of the capital, he was a strong head who was declared a deserter when he returned three days late to his regiment after a leave. Verdict: 2 years of hard labor. He joined the air force thanks to his mechanical skills and obtained his first commendations, which led to the erasure of his conviction. But he still enjoyed his leave in the capital too much, to the point of being declared a deserter again for not having returned to his squadron in time. This time, no punishment was given to the hero who fell in battle after his 10th certified victory.

6 - The price of glory

However, these exceptional privileges came at a price, which was paid in blood. According to the latest studies carried out by historian Lucien Morareau, out of 16,898 pilots licensed during the war, 2,221 (13%) died before the armistice from all causes, but only 916 pilots (5.5%) were killed in action, the rest being mostly lost in flight accidents.

For the 184 aviation aces, the losses in combat were almost four times higher, with 37 pilots falling under enemy bullets, i.e. a loss rate of 20%.

7 aces were killed in accidents during the conflict, four of them (Mathieu Tenant de la Tour, Marius Hasdenteufel, André Petit-Delchet, and Gilbert Deguingand) being victims of their recklessness in doing acrobatics to impress the public. The Spanish flu or lung diseases contracted during the war would take 5 others shortly after the end of the fighting. Two others were victims of psychological problems caused by the war from which they would never recover – madness following a concussion as a result of a crash for Joseph de Bonnefoy, and drugs for the young Constant Soulier who died prematurely in 1933.

Fighting on the front line was no picnic and physically demanding, consisting of two to three missions a day, weather permitting, at altitudes of up to 6,000 meters without oxygen masks. Ace Louis Risacher confirms this, "*It was very trying, and despite the fact that I was very athletic, I was doping myself by giving myself injections.*" Despite Heurtaux's testimony that the bar at N 3 Squadron was very sober and limited to a glass of Porto in the evening, this does not seem to have been the case at many other squadrons. The ace Marc Ambrogi, in the SPA 90 squadron, testifies to having taken a risky gamble, leaving the squadron's bar a little tipsy, to have narrowly escaped between the towers of the Toul cathedral. Fernand Chavannes testifies to the same fact for his squadron, the SPA 112, where alcohol flowed freely and where some pilots were more or less drunk while leaving on missions.

7 – Forever volunteers

For these pilots, doped up on adrenaline and fame, the post-war period was a difficult one, as illustrated by the quote from Roger Vercel's Captain Conan: "*Go ask a mutt to adapt to salad!* "Many of them, passionate about aviation, tried to make it their profession with varying degrees of success, some staying in the army, others working in aeronautical factories as test pilots, in flight schools, or in the first air transport companies. The fact that the Atlantic was to be conquered attracted several of them, including Charles Nungesser and Louis Coudouret, who lost their lives, Paul Tarascon, who eventually gave up, and Dieudonné Costes, the ace of the Eastern Front, who entered into eternal aeronautical glory by making the first Paris / New York flight between September 1st and 2nd 1930. But, in front of such a success, how many failures? 20 aces would perish at the controls of a plane during the inter-war period. Those who gave up aviation to become good fathers were only a minority... In 1934, the ace gunner Léon Vitalis took the initiative to create the association of the aces of the communiqué which would bring together these veterans in an annual dinner presided over by Fonck, and followed by journalists.

Above: Aviation journalist Jacques Mortane.

Twenty years later, when the Second World War broke out, 102 of the 175 French aces were still alive, taking into account some disappearances that had nothing to do with aviation. Now in their forties, they were no longer young enough to take part in front-line combat, but when reading their files, one is struck by the steps that most of them took to regain control of a fighter plane. Only a minority of 14 of them succeeded in doing so, those who had trained regularly in the reserves or in their profession, or those who benefited from high-level support, as was the case with André Dubonnet, who regained his place in escadrille SPA 3 thanks to the direct support of Paul Reynaud, President of the Council. If the reflexes and the science of piloting were blunted with age, this would not prevent three of them from adding one or more victories to their scoreboard. One of these aces, Hector Garaud, killed himself in training and another, Maurice Arnoux, fell in combat at the controls of his Morane 406.

The armistice was far from sending them back to their homes because more than a quarter of them (28) actively joined the Resistance or rallied Free France, a very high proportion. Thus, the ace Jean Gaillard de la Valdène was one of the first to join London as part of the small group of determined men who hijacked the trawler *Président Houduce* in Gibraltar. Wishing to fight at the controls of a Spitfire despite his 45 years of age, he went so far as to wear out the patience of General De Gaulle, who offered him several positions as military attaché abroad to represent Free France... Pierre Ducornet shared this desire to fight at the controls of a fighter plane, but understood that at his age he would be more useful in France, and in 1941 he committed himself to the Resistance, of which he became one of the heads of the civil and military organization in 1943. On the other hand, two aces of the First World War, Armand Pinsard and François Battesti, became active collaborators and were condemned by the justice system after the liberation, while two others, who became generals of the Vichy air force (Jean Romatet and Jean Jannekeyn) and were accused of state collaboration, were acquitted by the justice system but were disciplined by the air force by a compulsory retirement. Paul Gastin, another general who had accepted in 1943 the direction of the General Secretariat for Air Defense, was acquitted for acts of resistance.

After the liberation, the glory hour of the aces of 14–18, now in their fifties, passed from the attention of the media and the public, whose aviation enthusiasts devoured the stories of Pierre Clostermann's "*Le Grand Cirque*" (the Big Show) and other stories of the aviators of the Second World War. The 14–18 Aces Association would still meet until the dawn of the 1980's and it is in a great discretion that its last member, Louis Chartoire, would die on November 8, 1992 at the age of 97.

Fun facts

The ace of aces

This unofficial title, written by the journalist Jacques Mortane, went to the most successful pilot on duty at the front. It has been awarded successively to several pilots.

- Georges Guynemer from February 3, 1916 (5 v.) to April 3, 1916 (7 v.)
- Jean Navarre from April 3, 1916 (8 v.) to August 17, 1916 (12 v.)
- Georges Guynemer from August 17, 1916 (13 v.) to September 11, 1917 (53 v.) (killed in action)
- Charles Nungesser from September 11, 1917 (30 v.) to March 29, 1918 (31 v.)
- René Fonck from March 29, 1918 (32 v.) to the armistice (75 v.).

The youngest of the aces

Pierre Marinovitch was the youngest French ace: born on August 1, 1898, he was 20 years, 3 months and 11 days old at the armistice. He was 19 years, 5 months and 19 days old when he obtained his 5th victory. American pilot Putnam, born on December 10, 1898, was 19 years, 5 months and 23 days old when he became an ace.

The oldest of the aces

Adjduant Adolphe Du Bois d'Aische became an ace at the age of 43 years, 6 months and 5 days. The oldest ace pilot was Major Robert Massenet Royer de Marancour who obtained this title at the age of 37 years, 8 months and 4 days.

The greatest number of victories in one day

René Fonck obtained 6 victories in one day on May 9, 1918. He obtained 6 more on September 26, 1918. In both cases, these victories were obtained during two different missions carried out in the same day.

Two World Wars

As mentioned in the introduction, 14 aces of 14–18 flew operational flights during the French campaign in 1939–1940. Despite the fact that they were in their forties, this did not prevent three of them from claiming victories:

Marcel Haegelen won a certified victory on a Curtiss H-75 at the head of the Bourges defense patrol, against a Heinkel 111 bomber.

Marius Ambrogi in GC I/8 won a certified victory against a Dornier 17 in collaboration with other team members.

Lionel de Marmier claims 3 victories at the Polish GC I/145 (1 sure and 2 probable), but none will be confirmed.

It should be noted that two other pilots of 14–18 became aces in 1940:

Edouard Corniglion-Molinier added to his 2 victories won at the N 561 Venice squadron during the first war 4 others won at GC III/3 in 1940, aboard a Morane 406 and Curtiss H-75.

Robert Nast added to the 3 victories won at the 506 squadron in the East in 14–18 2 others won at the GC II/5 headquarters on Curtiss H-75, officially certified.

These two pilots will see their biographies detailed in the last volume of this encyclopedia in a section of special cases, which will also include the following four pilots:

Eugène Gilbert, a famous pre-war pilot, who became the first real French ace in 1915, before the army communiqué was instituted, but one his victories, however undeniable, was not officially confirmed.

Adolphe Pégoud, another famous pre-war pilot, some of whose victories were not officially recognized.

Gabriel Hébert, pilot of the Eastern Front, whose one victory was not confirmed for disciplinary reasons.

Pierre Malvoisin, a French maritime pilot, who did not formally certify the aerial victories of his pilots.

Readers' Notes

This encyclopedia presents short biographies of the French military aviation aces of the First World War, i.e. the pilots and gunners having obtained at least 5 aerial victories confirmed by the authorities and mentioned in le Journal Officiel de la République Française (JORF) - Official Journal of French Republic. 175 pilots and gunners of French nationality could claim this title, as well as 9 foreign pilots having fought in French squadrons and having won 5 victories in the French air force, that is to say a total of 184 aces. They are presented in alphabetical form; 6 particular cases are also presented in the last volume, for a total of 190 biographies.

At the beginning of each biography are listed the French decorations awarded to the pilot, the Legion of Honor, the Military Medal and the Croix de Guerre, the details of which were explained in the introduction of the first volume. It should be noted that foreign decorations are not included, nor the promotions in the legion of honor which were granted after the war.

The table of victories of each ace mentions the decisions that confirm the Confirmation of the victory by the French authorities. The proof comes in the form of an award, in most cases a citation in the army order, which is published in the JORF.

The date and type of the citation is given in the column "Confirmation" of the victories lists provided in every biographies:

"Citation OA" for citation to the order of the army, entitling to a palm on the war cross. These citations are published in the Journal Officiel (JORF), the date of publication and the page number of which are given when they are known. The citation may also grant the Legion of Honor or the Military Medal. It is then noted " Citation OA (LH) " or " Citation OA (MM) ".

"Citation OCA" for citation to the Order of the Corps, giving the right to a vermeil star.

"Citation OD" citation for a Divisional Order citation, entitling the recipient to a silver star.

Thus, for example, a citation in the Army Order awarded on August 21, 1918 and published in the Journal Officiel of December 11, 1918 on page 10 654, will be noted: "Citation OA 21-08-1918 - JORF 11-12-1918 p.10 654."

It should be noted that it is possible to consult all the citations to the order of the army mentioned in this encyclopedia on the following website :

https://gallica.bnf.fr/ark:/12148/cb34378481r/date.r=journal+official+of+the+r%C3%A9public+fran%C3%A7aise.langFR

Equivalences of French and American ranks

Equivalences of ranks between the French army of 1914 and the American army. Note that for the cavalry and artillery, the terms "*maréchal des logis*" and "*brigadier*" are used instead of sergeant and corporal.

Général d'armée	General
Général de corps d'armée	Lieutenant-General
Général de brigade	Major-General
Général de division	Brigadier-General
Colonel (Col)	Colonel
Lieutenant-colonel (Lt-col)	Lieutenant-Colonel
Commandant (Cdt)	Major
Capitaine (Cpt)	Captain
Lieutenant (Lt)	First Lieutenant
Sous-lieutenant (S/Lt)	Second Lieutenant
Aspirant (Asp)	Officer Cadet
Adjudant-chef (Adj-ch)	Chef Warrant Officer
Adjudant (Adj)	Warrant Officer
Sergent (Sgt) or Maréchal des Logis (MdL)	Sergeant
Caporal (Cprl) or Brigadier (Brig)	Corporal
Soldat (Sol)	Private

Albert Achard

First name: Albert.
Born: 26-03-1894 in Briancon (Hautes-Alpes).
Father's profession: Army officer.
Military pilot brevet: 20-04-1916 (N° 4642).
Units: MS 48, N 91 S, MF 98 T, F 203, N 510, N 97, N 85, N78.
Army communiqué: No.
Total score: 6 victories, 2 probables.
Died: 21-08-1972 in Briancon (Hautes-Alpes).

Left: Lieutenant Albert Achard standing near a Nieuport fighter.

Son of a soldier

Albert Achard was born on March 26, 1894 in Briançon (Hautes-Alpes), the garrison town of the 159th infantry regiment where his father served as a second lieutenant - two captains of the unit signed as witnesses on his birth certificate.

The young boy grew up and moved with his father's postings, while the family grew: a brother, Henri, was born in 1897, followed by a sister (Marguerite) in 1899 and a last brother (Jean) in 1909, when he was 15 years old. By this time, he had left the family home, which was located near Lyon: his father followed closely the schooling of his children, who all obtained good results and went to high school. The young Albert, who received a scholarship, was sent to the Prytanée Militaire de la Flèche as a boarder. This education allowed him to obtain both parts of the baccalaureate as well as a good knowledge of languages since he read English, German, and Spanish fluently.

Student at St-Cyr

At the age of 19, he followed his father's path by taking and passing the entrance exam for the St-Cyr officers' school in the summer of 1913. His father, whose only fortune was his modest captain's pay, obtained a scholarship (with trousseau) from the Ministry of War.

He officially entered the school on October 11, 1913 and began his schooling with a year of classes, a novelty because previously the students were immediately sent for a year of training in the troops. The First World War broke out while this first year was still in progress: the students of the class, known as the "Cross and Flag" class, were to be appointed second lieutenants and sent to regiments to serve on the front. On October 21, 1914, Albert Achard found himself a platoon leader in the 9th Hussar Regiment and thus participated in the race to the sea in the Somme, then in Flanders. His commanding officer, Colonel Burette, found him "*very young, incompletely trained, has everything to learn*".

Observer in the air force

However, he only stayed five months at the front with this regiment. As the cavalry was becoming obsolete in the trench warfare that was becoming widespread, he was sent on March 16, 1915, to the Le Bourget aviation school, to do an observer's

Right: Albert Achard in the cockpit of a Nieuport 17.

course.

Two weeks later, on April 1, 1915, he was sent to a squadron, MS 48, which was stationed in Lunéville in Lorraine. Equipped with Morane Parasols, this squadron was in charge of both major reconnaissance missions for the *Détachement Armé de Lorraine* (DAL) and fighter missions in this sector.

It is in this precise task that Second Lieutenant Achard would distinguish himself. On April 15, 1915, he was flying with MdL Chatain as a Morane pilot. The two men came across a German two-seater, class C, armed with a flexible machine gun operated by the observer. The Morane was armed only with the musket handled by Achard... but this did not prevent him from firing with skill. If the Morane had a hole in its wing from the German aircraft, the latter was hit by several rounds from the musket and was forced to land in its lines. The victory was awarded to the two men, who received a citation in the army order, which also mentioned for Achard an important reconnaissance carried out on 28 April.

A battle he fought a few weeks later proved to be even more perilous, if we are to believe the citation he received on July 1, 1915 in the Lorraine Army Order: "*During a barrage flight, armed only with a musket, he attacked an Albatross armed with a machine gun. Approached it at 50 meters, taking fire from three machine-gun bands whose bullets pierced the wings at the pilot's head and cut a post of the upper hut. Despite the danger of an imminent break-up of the aircraft, he continued the attack, firing 84 rounds at the enemy and pursuing him well beyond our lines until the aircraft suddenly nosedived. Thanks to the pilot's skill, he was able to land without accident with a completely damaged aircraft.*" The Morane Parasol's high wing was held up by only four masts and shrouds. The pilot maneuvered on eggshells to return to his base...

In the Eastern front in Salonika

A courageous observer, Albert Achard, promoted to lieutenant at the end of September, also proved to be a difficult officer. His superior notes "*Still very young, needs a firm and intelligent direction to bring out his excellent qualities.*"

This particular character is perhaps the explanation for his affectation to the Serbian army, facilitated by a squadron leader who only wished to get rid of him: he was officially seconded there on September 26, 1915, and embarked in Marseille on October 24, 1915, to arrive in Salonika on November 1. He was immediately assigned to the N 91 S squadron, equipped with two-seat Nieuport 10s, where he served as an observer.

Above: The black panther, symbol of escadrille SPA 78 where Achard finished the war.

He served there for several months, and won a new citation in the order of the Eastern Army on February 1st 1916, which indicated a long reconnaissance of 350 km that he accomplished all along the Vardar valley. Apart from reconnaissance missions, which he accomplished in a two-seater as a passenger, he found a passion: flying, learning "on the job" to maneuver a Nieuport. Although he did not attend any flying school, he obtained his military pilot's license n°4642 on April 20, 1916, awarded by the Eastern Army.

He was then assigned as a pilot to the MF 98 T squadron on April 20, 1916, in which he served for another three months. Major Denain, who was in charge of the local aviation, noted "*Very good military qualities. Still very young, pretentious. Has rendered real services as an army observer in the East. Passed his pilot's license exams with flying colors at the front. Will make an excellent pilot of fast aircraft. To guide and hold.*"

Return to France as fighter pilot

On July 22, 1916, Lt. Achard returned to France and joined the F 203 squadron as a pilot on August 18. But flying a Farman reconnaissance plane did not seem to suit his temperament: on January 1, 1917, he was sent to the Pau acrobatic school to learn how to handle fighter planes. He returned a few weeks later to be sent to the *Groupement des Divisions d'Entrainement* (GDE) in Plessis-Belleville, from where he was to be sent to an operational fighter unit.

However, he was not sent to a squadron, but to the N 510 fighter detachment of which he took command on March 10, 1917. These detachments were composed of four Nieuport 23 or 24 fighters operated by four pilots, including one officer, and were attached to an observation squadron whose aircraft they had to escort. The N 510 was thus attached to the F 44 squadron stationed at Froidos in Champagne. Albert Achard won a certified victory on May 2, 1917, by forcing an enemy two-seater of Schusta 7 to land in the French lines at Taon-les-Vosges, whose crew was captured.

The N 510 detachment was quickly disbanded and merged to form fighter squadrons. Achard and his three pilots found themselves assigned to the N 97 squadron on June 16, 1917. However, they kept their flying habits, because on June 27, 1917 Achard, MdL Mignot and Corporal Stenger, all former members of N 510, shot down a German plane over Sommedieu.

Removed from the command of N 85

This third certified victory earned Second Lieutenant Achard the Knight's Cross of the Legion of Honour. The general staff immediately transferred this young active officer to the command of a fighter squadron, N 85 stationed at Vadelaincourt, which he joined on June 29, 1917.

But this new position did not go well. Two and a half months later, he was relieved of his command by Major Voisin, head of aviation of the 2nd Army. Voisin explained this in the note he wrote for his subordinate: "*Has shown that he does not yet have the maturity of mind necessary for command, and, what is more serious, has a full and proud character which prevents him from listening to the advice he needs. Had to be relieved of his command as a result of punishment for lack of authority and unwillingness to comply with orders given to him.*"

Ace of Panthers

Following this failure, he was transferred as a simple pilot on September 12, 1917 to the N 78 squadron, called the Panther squadron, founded by Captain Pinsard but at the time commanded by Captain Eugène Verdon.

The young officer, detached from any command responsibility, matured and integrated quite well, being noted as "*Excellent officer, excellent pilot, bold and skillful fighter*". However, he is noted as having been ill a lot during the first half of

Albert Achard Victory List								
N°	Date	Unit	Plane piloted	Plane claimed	Place	Notes	German casualties	Confirmation
1	15-Apr-15	MS 48	Morane Parasol	Albatros C		As observer MdL Chatain as pilot		Citation OA 4-5-1915
2	2-May-17	Dét. N 510		Two-seater	Taon-les-Vosges		SchSta 7, Vfw Seifert and Uffz Wilhelm Niess captured	Citation OA 31-5-1917 - JORF 15 Sept 1917, p.7285
3	27-Jun-17	N 97		Enemy plane	Sommedieu	With MdL Mignot and corporal Stenger		Citation OA - JORF 11-9-1917, p.7141 - "2nd victory"
P1	10-Apr-18	SPA 78	SPAD	Enemy plane	E. Montdidier	With MdL Pavlovsky		
4	19-Jul-18	SPA 78	SPAD	Fighter	Crécy-au-Mont			Citation OA 2-8-1918 ("4th victory")
5	24-Jul-18	SPA 78	SPAD	Two-seater	Crécy-au-Mont			Citation OA 25 7 1918 ("5th victory")
6	26-Jul-18	SPA 78	SPAD	Triplace	Soissons			Citation OA - JORF 26-5-1919, p. 5423
P2	26-Jul-18	SPA 78	SPAD	Fighter	Soissons			Citation OA - JORF 26-5-1919, p. 5423 "probably shot down"

1918, perhaps due to malaria brought back from his stay in the Orient. The accounting book of SPA 78 for the second quarter of the year 1918 (the previous ones having disappeared) shows us that he had to be hospitalized in Clermont from May 2 to June 7, 1918. Before that, on April 10, 1918, he won a non-confirmed victory against a plane that came down near Mondidier with a crew member.

But these health problems seemed to be over in the summer of 1918, when the fighting against the German air force intensified and the Allied armies launched their counter-offensives that would lead to victory. In a few days, Achard won a whole series of victories. On July 19, 1918, he shot down an enemy fighter over Crécy-au-Mont, north of Soissons. Five days later, at the same location, he shot down a two-seater; then on 26 July he shot down a bomber he had been pursuing in the German lines. But he was attacked by a whole patrol of fighters: he probably shot down one of them in the ensuing melee and managed to return to his lines riddled with bullets.

Squadron leader at SPA 150

With six certified victories, he was now an "ace" but he did not get any glory from it, because the communiqué to the armies at that time only mentioned pilots who had reached the threshold of 10 air victories. These exploits attracted the attention of the head of the Groupement de Combat n°16, to which the SPA 78 squadron was attached. On August 27, 1918, he appointed Lieutenant Achard to lead another squadron of the group, SPA 150, a rather mediocre squadron if one believes its poor combat results. If Achard did not succeed in raising the score of this squadron under his command during the last months of the war, he did improve its performance and showed a sense of command that was unknown to him. The last assessment left on him by the head of the GC 16 is quite eloquent: "*During July and August, remembered the attention of his leaders by a very sustained work and several brilliant victories. Appointed to the command of a mediocre squadron, he made an excellent unit by his authority and example. Has kept intact its qualities of skill and bravery. Acquired the qualities of calm and command that he lacked. Valuable officer, very complete.*"

Promoted to temporary captain on October 25, 1918, SPA 150 under his command would win 3 certified victories (it had had 2 others before) but lose 3 pilots in combat. He remained at the head of the squadron after the armistice, and was transferred on May 10, 1919 to the head of SPA 69.

Little enthusiasm for an active career

He only stayed there for a short time: caught up in his past as a student at St-Cyr, he was recalled to this school on June 1, 1919 with the survivors of the so-called "cross and flag" class. He remained at the school for a while to attend classes, but then had to leave for the training school of his army and ended up in Saumur cavalry school on October 15, 1919.

He showed no interest in the training he was given: the airman had no interest in riding. The notes he received were scathing: "*Officer with a heavy and glorious war record. Does not have the excellent spirit of the officers of the "Cross of the Flag", his classmates. Also capable of doing very well or very badly. Not lacking in means, on the contrary, but from the beginning, completely disinterested in everything that was done at the school, both from a military point of view and in general education and riding. Sins through pride and weakness of character. Does not know how to dominate his nature.*"

He received a punishment of 30 days of rigorous arrest during his schooling and improved somewhat afterwards, enough to pass his exit exam without a single mention.

But, if his superiors in the cavalry were very severe on his behavior, it is necessary to note that those in the air force were of a completely different opinion. On April 16, 1920, after graduating from Saumur, he was appointed commander of a squadron of the 35th aviation regiment, where he gave full satisfaction: "*Commands the training section with firmness and competence. Excellent and bold pilot. Directed the training of 40 reserve pilots with safety, prudence, and authority.*"

Leaves the army to become a merchant in Albi

His private life took a turn at this time, as he married on June 1, 1920 Mrs. Marie Madeleine Battur, whose parents owned a clothing store on Rue Ste Cécile, opposite the Albi cathedral.

Captain Albert Achard quickly became interested in running his parents-in-law's business and took over the management. On August 15, 1921, while he was already the father of a child (two others were born in the following years), he obtained a 3-month leave of absence to take care of his business affairs.

He never returned to active service: on November 16, 1921, he was granted a 3-year leave of absence without pay to interrupt his seniority. At the end of this leave, his company was very prosperous and exported its articles of knitwear to Chile. Having become a well-known merchant in Albi, Albert Achard submitted his resignation from the army, which was accepted on 24 September 1924.

Remaining a reserve captain, he lost interest in the army and did not complete any reserve periods until 1930, when he flew one hour as an observer at the Istres school. He did a short period in 1933, then in 1938 he had to do 24 days of compulsory reserve. Operated on for a double hernia, he asked to be removed from the air force and to be transferred to the sedentary personnel - which was granted - but changed his mind after consulting his doctor.

He did not return to aircrew, but was assigned to a staff position where his superiors appreciated his general knowledge and noted his passion for teaching. He would be proposed, without follow-up, to the higher rank and to be promoted officer of the Legion of Honor.

On April 12, 1939, following international tension, he was called back to work until May 5. When war broke out, he was mobilized on September 2, 1939 and assigned to the intelligence center of the French Air Force in Carcassonne, where he was highly appreciated by his superiors. On January 9, 1940, he was assigned to the staff of the 2nd Air Division in Bordeaux.

He was demobilized even before the armistice came into effect on June 19, 1940, in the midst of the debacle. He then returned to Albi to find his business, and was struck off the Air Force reserve list on March 26, 1948, after having reached the age limit for his rank. He died on August 21, 1972 in his native town of Briançon.

Above: The airfield of SPA 78, Albert Achard's last unit.

Above: Morane L and Farman MF 11 of escadrille MF 98T in Dardanelles, 1915.

Above: Enlargement of the facing photo of airfield of SPA 78, Albert Achard's last unit.

Marc Ambrogi

First names: Marius, Jean, Paul, Elzéar – common name "Marc".
Born: 09-06-1896 in Marseilles (Bouches-du-Rhône).
Father's profession: Army administrative officer.
Military pilot brevet: 14-09-1916 (N° 933).
Units: N 507, N 90.
Army communiqué: 21-09-1918 (for 10 victories).
Total score: 14 victories (11 Drachen), 7 probables.
Died: 24-04-1971 in Marseilles (Bouches-du-Rhône).

Above: Marc Ambrogi near a Morane P reconnaissance plane.

Son of a soldier

Marius, Jean, Paul, Elzéar Ambrogi was born on June 9, 1896 in Marseille at his parents' home at 35 Place St Michel. His father Louis, 40 years old at his birth, was a Corsican from the commune of Ile-Rousse who worked as an administrative officer in the local military administration after having served in Tunisia.

The child, who was to take the customary first name of Marc, grew up in Marseille and probably moved to Montpellier as his father was transferred to the Hérault region in the early 1910s. His father ensured that his son received a thorough education and the young man obtained both parts of his baccalaureate and a good knowledge of the English language, which he spoke fluently.

Voluntary enlistment at 18

When the First World War broke out in August 1914, Marc Ambrogi had just celebrated his 18th birthday and had not yet done his military service, which normally takes place at the age of 20. Caught up in the patriotic fervor of the time, he voluntarily enlisted for the duration of the war at the town hall of the 2nd district of Marseille on September 25, 1914.

As a very young soldier, he was not sent to the trenches. Incorporated as a private on December 8, 1914 in the 14th section of the Administrative Clerks and Workers (COA), he became a cyclist in charge of carrying the staff folders to the various units. He cycled for almost a year: "*I didn't really do much. At 18, I was the youngest in the troop. I wasn't very strong and I couldn't carry the infantry bag, so I was appointed liaison soldier. We went around in Alsace, then to the Somme. But I never went into the dangerous places.*"

Eager to see action, he volunteered to serve in the air force several times in 1915. His request was accepted at the end of the year and on February 20, 1916, he entered the Dijon school for his theoretical training. On April 1, 1916, he was at the Juvisy flying school where he stayed for quite a long time (four and a half months) before obtaining his military pilot's license (n°933) on a Caudron G3 on September 14, 1916 - with two weeks of *arrêts de rigueur* (military punishment). A few days earlier, during the flying exams, he got lost in a cloud,

Above: Marc Ambrogi is sitting 2nd from the left with fellow pilots of the Cazaux gunnery school in December 1916. Sitting 1st from the left is the other future ace Basile Sauné, who won 5 victories in the Salonika front. (Emilie Durin collection)

Above: On July 20, 1917, Marc Ambrogi (2nd from right) is standing with fellow pilots from escadrille N 90, near a Nieuport 17 (note the peacock insignia on the fuselage and a SPAD VII in the foreground). 4th from the left is Lieutenant Pierre Weiss, commanding officer of N 90, who became a French Air Force general during WW2.

Above: 3rd from the left and a dog in the hands, Marc Ambrogi with other pilots from N 90.

and followed a railroad track... To end up in Orleans, where he ran out of gas in a maneuvering field. There, he managed to get some gasoline, and following a railroad track, reached the Chartres field as night was falling. He received a serious warning from his instructors... But he did get his license!

On September 15, 1916 he entered the Avord training school where he was put in command of the Nieuport fighter. On October 15, 1916, he received the rank of corporal, then on December 2, 1916, he went to the Cazaux shooting school, before finishing his training on December 26, 1916, at the Pau acrobatics school where he perfected his handling of fighter planes. On February 15, 1917, he was at the Groupement des Divisions d'Entrainement in Plessis-Belleville in the Paris region, awaiting his assignment to an operational unit.

Young fighter pilot on the Lorraine front

This was not long in coming: on March 3, 1917, he was assigned to the N 507 fighter section in Lorraine, a small detachment of 4 Nieuport fighters attached to an observation squadron to provide escort for its aircraft. This unit had a short life, since on April 10, 1917 it was merged with other detachments to form the N 90 squadron based in Toul, led by Captain Pierre Weiss.

The latter, an "old man" of 28 years, no longer had the enthusiasm he had at the beginning of the war and did not fly much. But he told his pilots, all of whom were his cadets, that he would give them medals if they brought home victories. The message was well received and the N90 included several pilots who became aces and shared the 35 victories the squadron won during the war: Charles Macé (12 victories), Maurice Bizot and Jean Pezon (10 victories each). However, there were several major obstacles on the way to glory: Lorraine was a sector considered to be calm, where German aviation was rare. The general staff was aware of this and did not entrust the best equipment to the fighter squadrons in the sector, which remained on Nieuport fighters for a long time.

This did not prevent Marius Ambrogi, who was appointed sergeant on May 10th 1917, from distinguishing himself by winning the first successes of his unit. On June 8 and 9, 1917, he strafed the enemy trenches with his Nieuport and forced an observation balloon to come down quickly, strafing it on the ground. This earned him his first decoration, a commendation in the divisional order which allowed him to wear the Croix de Guerre with a silver star.

The ribbon of this war cross was soon to be enriched with numerous palms. On July 3, 1917, in the company of Sergeant Bordes, he claimed a plane shot down near Nonsard but which was not confirmed, followed by two others on August 9 and

Right: Wearing all their decorations, Marc Ambrogi (right) and Charles Le Coq le Kerland, another ace from escadrille SPA 90 then SPA 82. Both aces shared a victory together on January 6, 1918.

September 11. For these actions, he received a citation in the army order on September 26, 1917, which, although it did not confirm the air victories, stated that two planes had been put out of action in the German lines on July 3 and September 11. This citation added a first palm to his war cross, and also earned him a promotion to the rank of *adjudant* (warrant officer) on October 10, 1917.

In all the brilliant actions of the squadron

Marius Ambrogi was also involved in all the brilliant actions of the N 90 squadron, and personally opened the official hunting list on October 30, 1917: in the company of Adjutant Bordes and two pilots from SPA 68, he forced a DFW C to land in the French lines, and it was captured with its crew at Commercy. He also continued his missions of strafing enemy positions: on November 20, 1917, he returned with his aircraft riddled with bullets.

All this fast-paced combat put the nerves of the 21-year-old pilot to the test. He held on with a good dose of Porto wine, which almost cost him his life, as he would testify in the 1970s in an interview recorded by General Barthélémy: "*After a victory or a hard fight, I returned to the field and, having been quite concussed by the fight, I was really shaken up. We had taken a few glasses of Porto. It was obvious that at that point we were drinking quite a bit. I made a stupid bet to pass with my baby Nieuport between the towers of the Toul cathedral. I took my plane – we were at the crossroads of Metz, not far from Toul – and some comrades went by car to check my exploit. I took*

Above: The rooster was the insignia of escadrille N 90.

altitude and went immediately to the cathedral. And I realized when I got to 100 or 150 meters that I would not be able to pass: I saw at that moment a lot of wires, which were probably used for the radio, which were coming from the towers of the cathedral. I found myself stuck. But I went ahead and passed on the edge, but a little bit close. I can assure you that I was completely sober that day. That little incident left us with a pretty bad memory."

On December 2, 1917, he received a new citation in the army order for his first official victory on October 30. He won a second official victory, which was also the 2nd of N90, on January 6, 1918 by shooting down a two-seater north of Regneville in collaboration with S/Lt Le Coq de Kerland. He was awarded the military medal for this on February 2, 1918, and promoted the next day to the rank of second lieutenant on a temporary basis.

The months of March and April saw other pilots of N90 inaugurate their hunting record by shooting down four certified and three probable aircraft. The Nieuports, struggling to catch up with the German observation aircraft, were then replaced by SPAD VIIs, which were delivered to the squadron along with a few Breguet 14s that were to form a reconnaissance section. It was probably on a SPAD VII that he won his third confirmed victory on May 16, 1918, against a two-seater of FA 242w shot down at Nomény in collaboration with Adjutant Dencausse.

Ace of the Drachen hunt

This was his last official success against an aircraft, because from the following day, May 17, he specialized in a completely different objective: hunting Drachens, the German tethered balloons used as observation platforms for artillery spotting. This was a military objective that was much easier to meet than the German planes, which were rarer in the skies over Lorraine, and also quite easy to shoot down in theory. But only in theory... Because the hydrogen-filled balloon, immobile in the sky, was invariably strongly defended by anti-aircraft artillery and even by protection planes. Any attempt to attack a Drachen leads the pilot to take high risks, as Marc Ambrogi himself will testify:

"*On SPAD, it was in 1918. When I returned from a mission I was alone, and I had not seen anything interesting. When I arrived above Dombasle, I think, I saw a sausage that was really tempting, a Drachen. I must have been at 3000 or 3500 meters, roughly. The sausage-drachen must have been at 1400. It didn't seem to be suspicious of me because there were a lot of clouds. I made a dive from 3500 to 1200 but the sausages were attached to extremely fast winches, and before I even got to this sausage it had been pulled up in a hurry and the two observers had parachuted out. I straightened up and the Drachen was almost back on the ground when I arrived. At the time of the resource, I had missed it, the Drachen on which I had fired had not ignited. I was 300–400 meters away and my engine started to sputter. I prepared myself for a hard fall, and completely panicked I had only one idea: to find a ground to land, my engine refusing to start again. As I put my wheels on the ground, I realized that my fuel pressure had dropped. This was a common phenomenon: when you went downhill a little too fast, the fuel pressure would drop. And we had the ability to turn a lever to pump. As my wheels were landing in a field, with machine guns firing at me, I was lucky enough to notice this. I pumped – completely panicked at the thought of being trapped. And my engine started up again, and I was able to get out of there, just as I was about to hit the ground. I came in low and I can tell you that the cold sweat was all over me.* "

He burned his first Drachen on May 17, 1918 in the company of MdL Jean Pezon, near Juville. Then a second one, his 5th official victory, on June 25, 1918 in the company of the same partner. It is in company of *adjudant* Maurice Bizot that he will set fire to his three following balloons, on July 30, August 7, and 21, followed by another one on September 2 set on fire with Pezon and Bizot.

This series of successes earned him new commendations in the army order, and on August 11, 1918, he was awarded the Knight's Cross of the Legion of Honor. Captain Weiss was surprised to see that all his young pilots seemed to be caught up in religious fervor, asking to be relieved of duty on Sunday mornings, in order to attend the 11 o'clock mass at the Nancy cathedral. They went there in their best uniforms and all their decorations, with the avowed aim of getting noticed by the young daughters of local notables, "*who were ogling us – we were young pilots!*" Captain Weiss, of Protestant faith, could only bow to this demand of his pilots to practice their faith. But he met with the bishop of Nancy, who confirmed that the different masses were all equivalent from the point of view of the Catholic Church, regardless of the time. He then summoned his pilots: "*Gentlemen, you will go to mass at 5 a.m. on Sunday. A light car of the squadron will be at your disposal!*" Needless to say, the religious fervor of the pilots decreased greatly after this measure...

Marc Ambrogi set fire to two other Drachen on September 15 and 16, in collaboration with *adjudant* Charles Macé and then *adjudant* Jean Pezon and Corporal Rivière. These were his 10th and 11th official victories, which earned him a place in the communiqué to the armies of 21 September 1918. This communiqué earned him fame because it was taken up by the national daily press, celebrating the arrival of a new ace of whom they published a short biography.

Above: Marc Ambrogi.

Confirmed in his rank of lieutenant on September 24, 1918, he increased his score on October 10 by burning a new Drachen, then two others on October 18 in collaboration with *adjudant* Macé for the first one, then Macé and Sergeant Auzuret for the second. He finished the war with 14 victories, the Legion of Honor, the military medal, and 10 palms on his war cross. These decorations were offered to Notre Dame de la Garde in Marseille, to thank the good mother for having survived the conflict.

An active reservist during the interwar period

After the armistice, Marius Ambrogi was still under the influence of his commitment and remained in his squadron stationed in occupied Germany. Obtaining a long leave of absence of one month on October 26, he carried out a convoy of German planes captured at the end of the year. SPA 90 was disbanded in the summer of 1919 and he himself was

Marc Ambrogi Victory List								
N°	Date	Unit	Plane piloted	Plane claimed	Place	Notes	German casualties	Confirmation
P1	3-Jul-17	N 90		Enemy plane	Nonsard	With Sgt Bordes		Citation OA 26-9 1917- JORF 29-12-1917, p.10 684 "out of control"
P2	19-Aug-17	N 90		Enemy plane				Citation OA 26-9-1917- JORF 29-12-1917, p.10 684 "falled vertically"
P3	11-Sep-17	N 90		Enemy plane	Varviney	With Sgt Bordes		Citation OA 26-9-1917 - JORF 29 12 1917, p.10 684 "out of control"
1	30-Oct-17	N 90		DFW C	Commercy	With Adj Bordes, Adj Dupré (N 68), MdL Planiol (N 68)	Capturé	Citation OA 18-12-1917 - JORF 17 1 1918, p.656
2	6-Jan-18	N 90		Two-seater	N. Regneville	With Lt Le Coq de Kerland		MM 2 2 1918
P4	27-Mar-18	N 90		Two-seater	Louvigny-sur-Seille	With Lt Le Coq de Kerland		MM 2-2-1918 "falling out of control"
3	16-May-18	SPA 90		Two-seater	Nomeny	With Lt Dencausse.	FA 242w, Ltn Gunther Amlinger & Ltn Robert Kamm, KIAs Raucourt 6 km from Nomeny	Citation OA 15-6-1918
4	17-May-18	SPA 90		Drachen	Juville	With MdL Jean Pezon.	Balloonzug 143.	Citation OA 15-6-1918 - JORF 16 Sept 1918, p.8127 "4th victory"
P5	21-May-18	SPA 90		Two-seater	Juville	With MdL Jean Pezon et Sgt Charles Macé		
5	25-Jun-18	SPA 90		Drachen	Goin-en-Moselle	With MdL Jean Pezon.	Balloonzug 152.	Citation OA 11-8-1918 (LO) - JORF 25 Oct 1918, p.9256 "5th victory"
6	30-Jul-18	SPA 90		Drachen	Grange-en-Haye	With Adj Maurice Bizot		Citation OA 25-8-1918 "6e victoire"
7	7-Aug-18	SPA 90		Drachen	Juville	With Adj Maurice Bizot		Citation OA 28-9-1918 - JORF 3 Jan 1919, p.292 "7th victory"

Mark Ambrogi Victory List								
N°	Date	Unit	Plane piloted	Plane claimed	Place	Notes	German casualties	Confirmation
8	21-Aug-18	SPA 90		Drachen	Vaxy	With Adj Maurice Bizot		Citation OA 28-9-1918 - JORF 3 Jan 1919, p.292 "8th victory"
9	2-Sep-18	SPA 90		Drachen	Juvelize	With MdL Jean Pezon and Adj Maurice Bizot.	Balloonzug 36.	Citation OA 9-11-1918 - JORF 20 Jan 1919, p.770 "9th victory"
P6	7-Sep-18	SPA 90		Two-seater	Baccarat	With S/Lt Dencausse and Adj Pezon		
P7	14-Sep-18	SPA 90		Enemy plane	German lines	With Lt Lemaire and Adj Macé.		
10	15-Sep-18	SPA 90		Drachen	Bourdonnaye	With Adj Charles Macé		Citation OA 9-11-1918 - JORF 20 Jaaaan 1919, p.770 "10th victory"
11	16-Sep-18	SPA 90		Drachen	Cirey	With Adj Jean Pezon and Cprl Rivière		Citation OA 9-7-1919 "11th victory"
12	10-Oct-18	SPA 90		Drachen	Amelecourt			Citation OA 9-7-1919 "12th victory"
13	18-Oct-18	SPA 90		Drachen	Omeney	With Adj Charles Macé		Citation OA 9 7 1919 "13th victory"
14	18-Oct-18	SPA 90		Drachen	Avricourt	With Adj Charles Macé and Sgt Auzuret		Citation OA 9-7-1919 "14th victory"
15	18-May-40	GC I/8		Dornier 17	Cambrai	With Lt Bouysse and Sgt-Ch Corrin		Citation OCA 5-6-1940

promoted to lieutenant and reassigned to the Istres flying school. It is there that he presented his resignation from the active army, which was effective on December 25, 1919.

From then on, he devoted himself to his private life and gained a good professional situation, creating a coal import business. On December 6, 1924, he got married and started a family with two children (a boy and a girl).

But in parallel to his professional and private life, he remained a reserve officer and was always interested in aviation, leading the Provence flying club of which he was president for forty years. Assigned to the reserve in the 32nd aviation regiment, he regularly performed voluntary reserve periods to keep up his flying skills, becoming a member of the regional air club of Marseille. The army showed interest in him by appointing him captain of the reserve on October 24, 1926, and officer of the Legion of Honor the year before. During the inter-war period, he added nearly 1000 hours of flight time to those obtained during the First World War.

In 1935, when the French Air Force classified its pilots by categories, he was classed as a "B" pilot, i.e. fit for combat on bombing or reconnaissance aircraft, and at the end of the year he was classed as an "A" pilot, fit for fighter aircraft. This is a rare privilege for a reservist in his forties and is a testament to his excellent training. Made Commander of the Legion of Honor that same year, he flew the following year on the Morane 225 or Dewoitine 501 fighter aircraft and was appointed Reserve commandant (Major) on December 25, 1936.

In combat during the Second World War

Even before the outbreak of the Second World War, he was recalled to the army on August 26, 1939 and assigned to GAO

Above: Bloch 152 fighter of commandant Marc Ambrogi in 1940. Note the personal decoration: Dopey playing flute…

515 (Aerial Group of Observation), undoubtedly the air unit closest to his home. He only spent a short time there, as he was assigned to the Montpellier Fighter Training Center on September 2, 1939.

His level of training allowed him to be assigned to an operational fighter group. He is then transferred on December 3rd 1939 to the Groupe de Chasse (Fighter Group) I/8, as second in command. The GC I/8 would use the phoney war to transform itself on the Bloch 152 fighter, which knows serious problems of development at the level of its propulsion group. Marc Ambrogi was given one (n°231, registered Y-718), which he decorated with a personal insignia on the vertical stabilizer, the dwarf Dopey playing the flute, and decorated with a tricolour "ace band" on the fuselage. He is indeed famous in the group for playing this musical instrument... His n°231, which was equipped with a Gnome and Rhône 14N25 engine, would be transferred to the Groupe de Chasse III/10 in April. He would not carry out any combat mission on board this aircraft because he receiveed n°522, equipped with a more powerful Gnome and Rhône 14N49, as a replacement aircraft.

The German attack took place on May 10th 1940 and the Panzerdivisionen passed through the Ardennes forest. Despite reports from the French reconnaissance planes, no immediate reaction from the air force took place to try to prevent them. It was only on May 13th 1940, when the German troops came out of the forest and tried to cross the Meuse river, that the French air force retaliated with all it had. The GC I/8, based near Nancy since April 1st 1940, moved on May 12th to La Fère (near Laon) to intervene in the battle.

The commander, Marc Ambrogi, would carry out a total of 4 war missions during the campaign, on his Bloch n°522. On May 14, he covered his ground without encountering any enemy. Then, in the late afternoon of May 15, 1940, he entered the fray by flying a mission over Montherlé with 5 other pilots, accompanied by fighters from other units, to escort French bombers attempting to bomb the bridges over the Meuse. He would testify in these terms:

"*The LeO 45 bombers, instead of coming in groups, came in single file. I don't know who was the officer who gave this stupid order, because the German flak was terrible. These bombers, in order to be sure to have their objective, had come to a maximum altitude of 100 meters. There were about 40 of fighters to fly above this, and we stayed almost to the end of our gasoline. I was sad to see two LeO 45s shattered by shells. There was a kind of mist... The flak was one of the strongest I have ever seen. There were about forty planes flying by each other at full speed, it was a complete panic. I remember that I got completely lost in the chaos, I didn't know where I was. I was lucky enough to see one of my teammates signal me to get back in our lines – I knew I was there because the flak was gone. I followed him, and as I got closer to him (phonics not working as usual) I signaled that I didn't know where I was. He told me the same thing.*

Nieuport 23 n°2782 of Escadrille N 90, Toul, late 1917

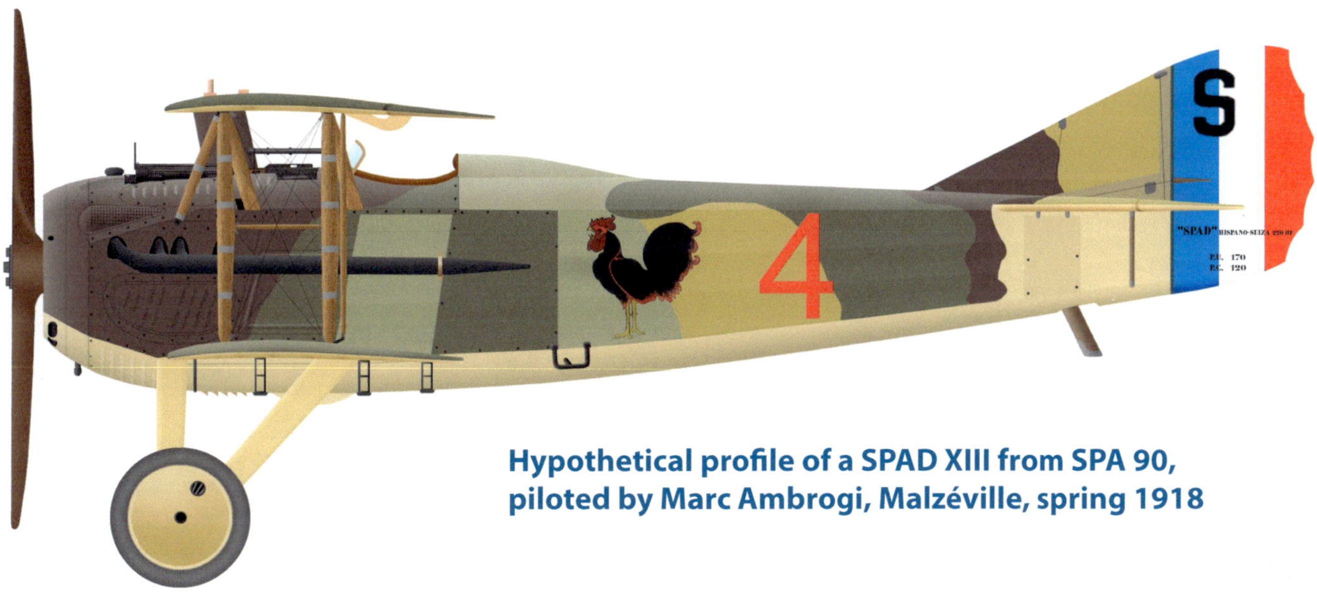

Hypothetical profile of a SPAD XIII from SPA 90, piloted by Marc Ambrogi, Malzéville, spring 1918

Bloch 152 n°231, GC I/8, commandant Marc Ambrogi early 1940. Ambrogi scored one victory in 1940, but on another Bloch 152

So he took a course, I was in his tail. We took a course to the southwest, sometimes fired upon by French flak. I was getting worried because my fuel level was getting lower and lower. At one point, we went down to 4 or 500 meters to look for our way, to look in the air to see if we could see any planes. Anyway! I see him approaching me, signaling that he is going to land. Landing like that in a field with a Bloch 152... He spotted one wide enough, narrow enough, and near a road. I turn around in slow motion to watch his landing. He missed the field the first time. The second time I see him succeed in his approach. At the moment he lands, a cloud of earth. He stays on his wheels, propeller spinning. I see my pilot, who was second lieutenant Thollon, get out of his Bloch, talk to a cyclist who was passing on the road, and get back into his plane, make his turn, and try to take off again. At that point, my gas was really low. I was wondering where to land before I ran out of gas, when I looked up and saw three Morane 406s passing by. I followed them without much throttle, thinking they knew where they were going. They had not flown for five minutes when, behind a hill, I saw an airfield: it was the airfield of Cambrai! I saw my three Moranes landing, and I saw one of them doing a magnificent barrel roll. The airfield had just been bombed. The first two landed, the third one went down on its nose. I was almost out of gas, I knew it was a French field. I had to try everything. How I landed, between two shell holes, I do not know. I returned to my squadron the next day, after having been received admirably by the commander of the fighter group. They were less fortunate than we were, for they lost two pilots in that affair of bombing the bridges of the Meuse."

On May 18, 1940, when the front was broken, Major Marc Ambrogi took part in the destruction of a Dornier 17 bomber over Cambrai, in collaboration with Lt Bouysse and Staff Sergeant Corrin. This 15th victory earned him a citation in the order of the fighter aircraft (war cross with vermeil star) which was awarded to him on June 5, 1940.

But on May 19, 1940, his Bloch 152 n°522 was destroyed in aerial combat while being piloted by Lt. Ruby, who was killed on board. Without a fighter, Maj. Ambrogi made himself useful by flying a small Salmson Phrygane liaison plane, a tourist plane that allowed him to ferry pilots from base to base during evacuations to the South. He would evacuate the Morane 406 personal of General Pinsard (n°1060) painted in black, that the pilots recovered on the field of Brétigny.

Just after the armistice, he was assigned on August 1, 1940 to the II/8 fighter group before being demobilized five days later. He then retired to Marseilles while the Vichy regime was being established in France.

Returning to his profession as a shipowner, Marc Ambrogi was undoubtedly an admirer of the personality of maréchal Pétain and joined the Legion of Combatants, a mass organization organized by the regime and intended to act as a relay for public opinion, relying on the vast majority of veterans of the First World War among whom Marshal Pétain was popular. He remained a member of this organization until 1944.

This would get him into trouble after the liberation. On August 13, 1945, he declared to the air force authorities that he had not been able to respond to General de Gaulle's appeal of July 8, 1944, enjoining all French officers, whether active or reserve, to take part in the Resistance battles to liberate the country. "*My very poor health at that time did not allow me to respond to General de Gaulle's appeal of July 8, 1944. But as soon as my health allowed me, I joined the resistance.*" He specified that he had taken part in the fighting for the liberation of Marseille (August 1944) in the Montolivet district (12th arrondissement) with the possibility of providing a certificate issued by the local resistance leader.

According to a note from the air force security service, he was prosecuted after the liberation for having been a member of the legion of combatants (which was the case for millions of French people), but also accused of having been a member of the militia, the ultra-collaborationist military organization. The Marseilles court of justice acquitted him of these charges in a judgment of November 14, 1945, rendered after additional information.

After the liberation, he resumed his functions as a company director (he was the general manager of the Fouquet-Sacop company, specialized in the armament of tankers), and also resumed his functions at the Provence flying club.

Appointed as a reserve Lt. Colonel on July 1, 1948, he became President of the national association of air force reserve officers (Marseille section) from 1950 to 1952, and then became its honorary president after that date. He was disbarred from the Air Force in 1957, having reached the age limit.

He was also president of the Bouches-du-Rhône "Vieilles Tiges" (old flyers) club and died in Marseille on April 24, 1971.

Right: Nieuport of Ambrogi, Escadrille 90.

Below: N 90 pilots and their mechanics. Marc Ambrogi is on the center, just right form Cpt Weiss wearing a clear colored kepi.

Paul D'Arguéeff

Name: Argueïev (Russian) / D'Arguéeff (French).
First names: Pavel, Vladimirovitch (Russian) / Paul (French).
Born: 1-3-1887 in Yalta (Russia).
Father's profession: Engineer in steamships.
Military pilot brevet: 30-1-1916 (n°2573).
Units: N 48, Russian 19th KAO, SPA 124 – GC 21.
Army communiqué: No.
Total score: 9 victories, 1 probable (15 victories and 2 probable total score).
Died: 30-10-1922, in Trutnov, (Czech Republic) – Killed in a flying accident.

Above: Portrait of Paul d'Arguéeff taken after the was with all his Russian and French decorations. The photo is dedicated to the Hispano-Suiza engine company: "Your engines who so many times saved my life and allowed me to win against the Boches are irreproachable. D'Arguéeff"

Pavel Vladimirovitch Argueïev (Павел Владимирович Арреев) was born in Yalta (Russia) on March 1, 1887 into a rather well-to-do Russian family whose father was an engineer specializing in steamships. The young boy received a military education by attending the military academy of Odessa in 1905 where he was appointed second lieutenant, and from which he graduated with the rank of lieutenant in 1907. Contrary to a persistent legend, he did not become a lieutenant-colonel and resigned from the Russian army for refusing to punish a soldier, but was sent to France on a military mission, settling with his wife on rue Mirabeau in the 16th district of Paris.

He was still there when the war broke out and received orders from his superiors to fight in the French army where he was integrated into the 131st infantry regiment with the rank of lieutenant on August 25, 1914, following a ministerial decree published in the official journal of August 30 where his name was changed to Paul d'Arguéeff. Wounded in the leg during the Battle of the Marne on September 23, 1914, he was given command of a company the following November and promoted to the rank of captain. He distinguished himself during the attack on Vauquois, for which he was commended in the army order on December 28, 1914. Wounded several times in combat, he was awarded the Knight's Cross of the Legion of Honor on May 2, 1915. Recognized as unfit for infantry combat due to his wounds, he volunteered to join the aeronautical service on July 22, 1915.

He then joined the flying schools from which he graduated on January 30, 1916 (n°2573) and was assigned to the N 48 squadron from May to August 1916, flying quite little. He then left France for Russia where he joined the 19th Korpusnoi Aviatsionniy Otryad (KAO) on October 20, 1916, led by the Russian ace Alexander Kozakov, and proved to be a particularly gifted fighter as he won 6 air victories between January and June 1917.

But the Bolsheviks' seizure of power in December 1917 made his position difficult as a former officer of the Tsarist

Above: Paul d'Arguéeff in front of his SPAD XIII, wearing number 19. Member of the staff of Groupe de Combat 21, he was attached to escadrille SPA 124 and is still wearing the rank of capitaine.

army. He then decided to leave the country and managed to reach France on April 4, 1918, where he was assigned to the SPA 124 squadron integrated into the GC 21, of which he became a staff officer. There he resumed fighting at his own pace, according to the testimony of his squadron mates, spending his time sleeping under the wing of his plane. When the ace comes out of his sleep... It was to attack the enemy and bring his SPAD to point-blank range to shoot him down with a burst of fire. He won 9 new victories, bringing his final score to 15, which earned him the rosette of Officer of the Legion of Honor in April 1919.

He did not wish to return to Russia after the armistice and found a job as a pilot in the Franco-Romanian company which flew to Eastern European countries. On October 30, 1922, his Potez, caught in the fog, hit a hill in Trutnov, in the Sudetenland region of Czechoslovakia, and he was killed instantly. His comrades of the SPA 124 squadron organized a collection to support his widow.

Left: 3rd from left and smoking a big cigar, commandant Paul d'Arguéeff with SPA 124's pilots.

Paul D'Arguéeff Victory List								
N°	Date	Unit	Plane piloted	Plane claimed	Place	Notes	German casualties	Confirmation
1	10-Jan-17	19e KAO	Nieuport	Albatros C				
P1	8-Apr-17	19e KAO	Nieuport	Fokker	Mitau		FA 286, Ltn Wilhelm Höding et Ltn Alfred Küffner killed.	
2	21-Apr-17	19e KAO	Nieuport	Biplace			FA 257, crew captured	

SPAD XIII n°4872 of commandant Paul d'Arguéeff, GC 21 (SPA 124), summer 1918.

Paul D'Arguéeff Victory List								
N°	Date	Unit	Plane piloted	Plane claimed	Place	Notes	German casualties	Confirmation
3	6-May-17	19e KAO	Nieuport	Hansa-Brandenburg C.I (26.51)	Brezezany	With Cpt Alexandr Kozakov and two other pilots		
4	17-May-17	19e KAO	Nieuport	LVG C	Jakobstadt	With Cpt Alexandr Kozakov		
5	8-Jun-17	19e KAO	Nieuport	Hansa-Brandenburg C.I (63.75)	Podgartze, near Kozova	With Cpt Alexandr Kozakov	Jasta 34b, Ltn Anton Warmuth killed in Filain	
6	20-Jun-17	19e KAO	Nieuport	Rumpler C.I (4739/16)	Tsmenitzi, near Nejnokov	With Cpt Alexandr Kozakov		
7	1-Jun-18	SPA 124	SPAD XIII	LVG C				Citation OA - JORF 9-12-1918 p.10598
8	13-Jun-18	SPA 124	SPAD XIII	Rumpler C	Puisieux-Beaumont			Citation OA - JORF 9-12-1918 p.10598
9	14-Jun-18	SPA 124	SPAD XIII	Biplace				Citation OA - JORF 28-11-1918 p.10271
10	26-Jun-18	SPA 124	SPAD XIII	Biplace				Citation OA - JORF 6-12-1918 p.10531 "4rd victory on the french front"
11	27-Sep-18	SPA 124	SPAD XIII	Fokker D.VII	N. Cerny			Citation OA - JORF 11-1-1919 p.392 "3 new german planes"
12	28-Sep-18	SPA 124	SPAD XIII	Biplace	Sechault			Citation OA - JORF 11-1919 p.392 "3 new german planes"
13	28-Sep-18	SPA 124	SPAD XIII	Biplace	Sechault Laval			Citation OA - JORF 11-1-1919 p.392 "3 new german planes"
P2	5-Oct-18	SPA 124	SPAD XIII	Biplace	NE. Autry			
14	5-Oct-18	SPA 124	SPAD XIII	Pfalz D	Orfeuil			
15	30-Oct-18	SPA 124	SPAD XIII	Biplace	E. Quatre Champs			Officier LO - JORF 9-4-1919 p.3719 "shot down 9 new German planes [in France], raising to 15 the number of his victories"

Maurice Arnoux

First names: Maurice, Albert, Alfred, Jean.
Born: 07-09-1895 in Montrouge (Hauts-de-Seine).
Father's profession: High-ranking civil servant.
Military pilot brevet: 24-05-1916 (N° 3539).
Units: MF 99 S, MF 55, N 94.
Army communiqué: No.
Total score: 5 victories, 1 probable.
Died: 06-06-1940 in Angivillers (Oise) – Killed in aerial combat.

Above: Maurice Arnoux during WW1.

Young man with a passion for aviation

Maurice Arnoux was born on September 7, 1895 in Montrouge. His father, Eugène, was a deputy head of office at the Conseil d'Etat (French supreme administrative court of justice) and in 1899 he wrote a book on litigation before this court.

The young Maurice does not seem to have a particular passion for law. As a student at the Lycée Michelet in Vanves, he got into the habit of going to the airfields after class to watch the pioneers fly their fragile machines. He continued his studies at the electricity and industrial mechanic school to become an engineer and, in 1912, obtained his first flight thanks to Pierre-Émile Gougenheim, chief pilot at the École Farman in Étampes. He developed a passion for mechanics and tinkered with small airplane engines with a friend whose father owned a mechanical workshop in Montrouge.

Adventures in Serbia 1915

When the Great War broke out, he was mobilized with the class of 1915 in December 1914 and joined the 2nd aviation group in Lyon-Bron, from where several squadrons were formed. He was then assigned, as a private 2nd class employed as a car driver, to the MF 99 S escadrille which had to leave to support Serbia. The men left by boat from Marseille on March 3, 1915 and stopped over in Malta and Piraeus before reaching Salonika, from where they left by train for Belgrade, which was reached on March 16, 1915. The MF 99 S squadron settled in the nearby Banjica airfield in rudimentary conditions and the first flight took place on April 23. This was followed by five months of successful aerial reconnaissance missions on the Austrian side of the Danube. Maurice Arnoux, a simple soldier, did not make any operational flights and was employed for ground tasks. But the offensive led by the German troops in October 1915 was to break through the front and force the MF 99 S to withdraw with the Serbian army to the south of the country. Then began a long and delicate journey at the controls of a truck on the sinuous and snowy roads of Serbia, which ended in the city of Prizen on November 17, 1915 where the Serbian army held its last square. The French had to send their planes by air to the town of Schköder, while the rest of the personnel, leaving their vehicles and heavy equipment behind, had to leave in a column on foot through the Albanian

Right: A Berliet truck such as Maurice Arnoux was driving, during the campaign of escadrille MF 99 S in Serbia, 1915. (Boris Ciglic collection)

mountains. Maurice Arnoux managed to reach the Albanian coast with the other mechanics despite the terrible weather conditions and was finally evacuated to Italy, then France, on Christmas Day 1915.

Military pilot

Like all the men of the MF 99 S squadron, he received a Serbian decoration, and had to spend a few months in hospitals in the Lyon region to rest after this trying journey. As soon as he was discharged, he asked to go to a flying school, which was granted. He then followed the cursus of the flying schools, gaining Dijon, Avord, Etampes and then Chateauroux, which he left on August 1, 1916 with his military pilot's license n°3539 and promoted to the rank of corporal.

After spending some time at the GDE in the operational reserve, he was assigned to the MF 55 squadron on Maurice Farman observation aircraft, which was operating in the Verdun region at the time of the violent fighting to retake the Douaumont fort. Maurice Arnoux was promoted to Sergeant on October 27, 1916. He flew numerous reconnaissance flights in the area and was cited in the Army Corps Order on November 10, 1916.

On March 21, 1917, he returned to the GDE and received a new assignment to a fighter squadron, which was to be N 49 operating on the Alsace front in the Belfort region, which he joined on April 16, 1917, and where he remained until the end of the war. After a difficult start, where he was shot down without damage on two occasions, he won his first confirmed victory in collaboration with two other pilots on June 26, 1917 against a plane shot down in the town of Hirsingen. Promoted to warrant officer, he attacked German captive balloons, the Drachen, and set fire to two of them in June 1918, then closed his hunting record by destroying a Rumpler C in collaboration with two other pilots on October 7, 1918 at Burnhaupt, then another two-seater on October 21 over Mulhouse.

He finished the war with 5 certified victories and one probable victory, and was awarded the military medal on October 4, 1918.

Civilian pilot during the interwar period

He remained in the army for some time after the armistice, being transferred to SPA 91 on April 9, 1919, which he left five days later due to its dissolution. He returned to civilian life and was awarded the Legion of Honor on July 13, 1919.

He then set up a saw blade and sawing machine manufacturing business in Montrouge with childhood friends, which gradually enabled him to earn a very decent living. At the same time, regularly doing his reserve periods, he kept his hand in piloting and was promoted to second lieutenant, then reserve lieutenant in March 1922, captain in 1930, and commander in 1938.

When his personal fortune allowed him to do so from 1930 onwards, he started flying in sports aviation and took part in several competitions. He finished second in the Dunlop and Zenith cups in 1931. In 1933, he beat the 100 km speed record for a 350 kg single-seater at an average speed of 304 km/h. He becomes famous in the sporting world by winning the Suzanne Deutsch de la Meurthe cup in 1934 on board his Caudron-Renault C.460, covering the 2000 km of the circuit at an average speed of 389 km/h. He was at the climax of his sporting career in 1935, by winning the 3rd place of the cup Deutsch de la Meurthe but by beating a world record on 100 km in his category, and by winning several competitions such as the 12 hours of Angers, the Deauville-Cannes-Deauville, the Zenith cup, and the Michelin cup.

Falling in combat in 1940

In 1939, he was elected Mayor of Chamarande, in the Essonne, where his family had a house, and also became a local representative. He was mobilized again with the rank of commandant (major): with an excellent level of piloting, he was classified as a category A pilot and could claim to serve in a fighter group. He was then entrusted with the second-in-command of GC III/7, a group of reservists flying Morane 406s, an outdated aircraft facing the enemy's Messerschmitt fighters. Although Major Arnoux was a dynamic leader, it was clear that at the age of 45, he was not as quick to recover from the fatigue of flying as his pilots, who were 10 to 20 years younger than he was, as his deputy Captain Bouvarre testified.

On May 10, 1940, when the German attack began, the GC III/7 airfield was bombed at dawn by surprise and 3 Moranes were damaged. Maurice Arnoux then took off at the head of a patrol of three, and managed to intercept a Heinkel 111 on its way back, between Châlons-en-Champagne and Vitry. He made several firing passes, but was hit by the bomber's gunner who wounded him with a windshield fragment in the eye and also hit the engine. Arnoux had to break off the fight while drifting in white smoke; he decided to land his Morane (n°813, coded 6, borrowed from S/Lt Challe) in a hurry and overturned on landing in the commune of Couvrot, one of his tires having been punctured by the enemy fire.

Resuming its flights very quickly, it fell in aerial combat on June 6, 1940. At the head of a double patrol of 6 aircraft, quickly reduced to 5 due to a mechanical problem, he encountered a group of Bf 109s flying at a higher altitude. While a lone Bf 109 tried to attack a Morane, others dived

Above: Maurice Arnoux as a racer pilot during the 1930's. Note the Caudron race in the foreground.

Above: Maurice Arnoux in with uniform during WW2, with the rank of commandant.

N°	Date	Unit	Plane piloted	Plane claimed	Place	Notes	German casualties	Confirmation
1	6/26/17	N 49		Enemy plane	Hirsingen	With MdL Hapedanne de Belville.		Citation OCA
P1	9/17/17	N 49		Enemy plane	Aspach	With Sgt Butcher.		
2	6/8/18	SPA 49		Drachen	Tagsdorff			Citation OA - JORF July 23, 1918 p.6346
3	6/26/18	SPA 49		Drachen	Ferrette			Citation OA - JORF Aug.26, 1918 p.7537
4	10/7/18	SPA 49		Rumpler C	Burnhaupt	With MdL Girard and MdL Simonin.		Citation OA - JORF Jan.11, 1919 p.508 "4th victory"
5	10/21/18	SPA 49		Two-seater	Mulhouse	With S/Lt Bouyer, S/Lt Brétillon and Adj Hamot.		Citation OA - JORF Feb.3, 1919 p 1268 "5th victory"

Table: Maurice Arnoux Victory List

on the French fighters. One passed in front of the major's Morane. Maurice Arnoux dived to shoot the Messerschmitt, but he did not see a third Bf 109 patrol coming from behind, despite the radio call from his crewmate, Captain André Houpeurt. His Morane 406 n°132 coded "2" (N 452) was shot down and crashed near the road 36 near Angivillers in the Oise region, 15 km north-east of Clermont.

Left: Maurice Arnoux with a Morane 406

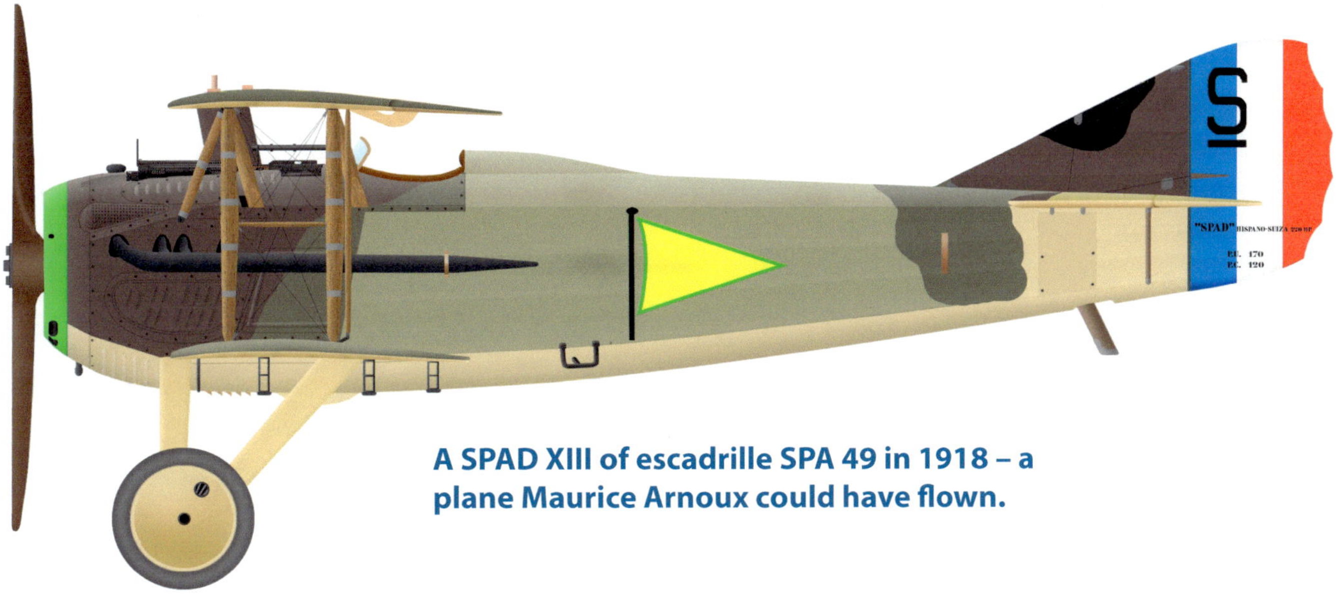

A SPAD XIII of escadrille SPA 49 in 1918 – a plane Maurice Arnoux could have flown.

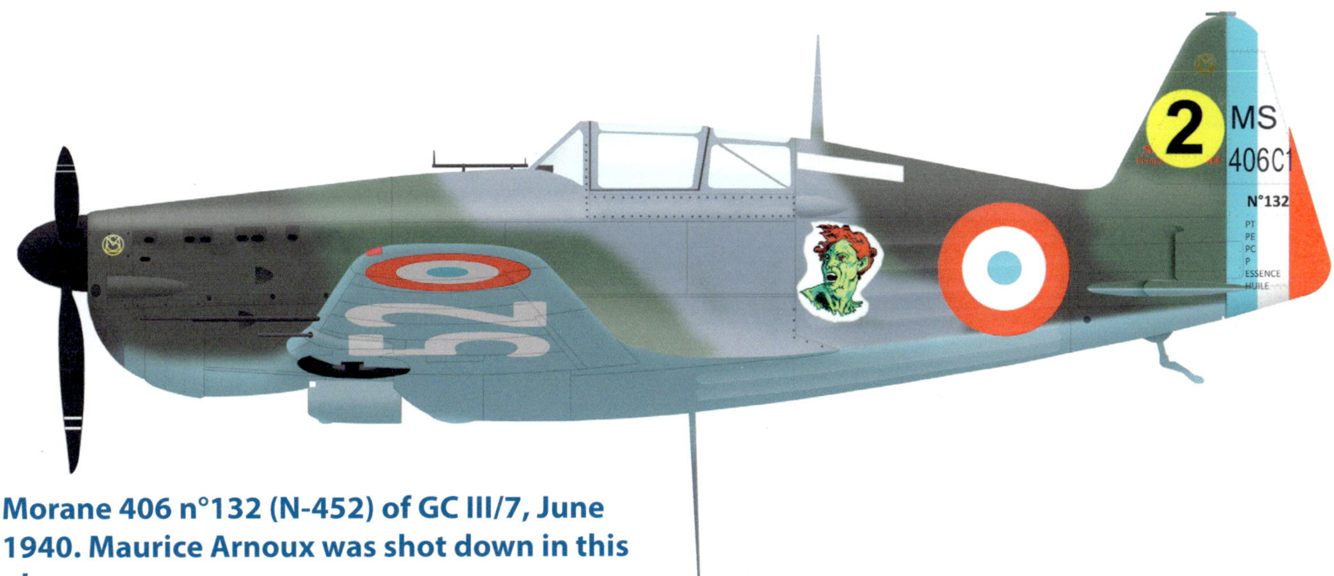

Morane 406 n°132 (N-452) of GC III/7, June 1940. Maurice Arnoux was shot down in this plane.

Jean Arpheuil

First names: Jean, Henri, Maurice.
Born: 06-10-1892 in Montluçon (Allier).
Father's profession: Butler.
Military pilot brevet: 08-04-1917 (n°5907).
Units: SPA 151, SPA 170.
Army communiqué: No.
Total score: 5 victories, 1 probable.
Died: 20-06-1920 in Cambo-les-Bains (Pyrénées-Atlantiques) from disease (Spanish influenza complications).

Above: Jean Arpheuil portrait.

The whole life of Jean Henri Maurice Arpheuil was a tragedy. He was born on October 6, 1892 in Montluçon, in the Allier region, where his father was the butler of the railway station hôtel, and passionate about sports because he became president of the local Sport Club.

Growing up, Jean became a young man of 1m78 rather tall for the time to whom his father communicated his passion for sports since he participated with his brother Maurice in several local cycling competitions. The two brothers did not just make up the numbers, since Jean became the champion of the Allier region, and Maurice came second in the French championships held in Nantes in 1910. Unfortunately, two tragedies hit Jean when he was not yet 20 years old: his brother Maurice died in October 1911, followed by his father in July 1912. He then inherited his father's small business because he was declared to be working as a butler before being drafted on October 8, 1913.

He was incorporated into the 121st Infantry Regiment as a private. Well noted by his superiors, he was promoted to the rank of corporal on February 17, 1914 and became a bicycle scout, finishing second in a military competition (the Wolber competition) organized in July 1914. War broke out the following month and Corporal Arpheuil immediately took part in the first engagements. His qualities as a leader of men were quickly revealed as he was quickly promoted to the rank of sergeant on September 2, 1914, a month in which his regiment was particularly tested, and was then appointed second lieutenant on a temporary basis on October 15.

On January 5, 1915, he led a reconnaissance operation in the commune of Saint Aurin in the Somme. An enemy shot hit him in the face and back with an explosive bullet. He was then evacuated to the Breteuil hospital and left on April 28, 1915 to convalesce, keeping from this misadventure a scar on his left cheek, but also his first mention in dispatch (in the regimental order) obtained on January 8, 1915: "*Was wounded during small operations where he showed serious qualities of courage and coolness*".

Back to his regiment on July 8th, he left the next day to do a training course as a machine-gunner until August 4th when he returned to the front to lead his machine-gun section. He fought relentlessly until the end of 1915 and suffered a second bullet wound on December 14, 1915, when he was superficially hit in the back in the Armancourt sector. Treated on the spot, he was withdrawn from the front on January 16, 1916 to follow a new training course at the Cemachine-ginners instruction

Above: SPA 151 pilots in Clermont-de-l'Oise, summer 1918. From left to right: Lt Arpheuil, Lt Henri Hirsch, unknown, Lt Philippe Koechlin-Schwartz. Hirsch scored 3 victories, including Zeppelin L 50 (LZ-89) captured in Montigny-le-Roi on October 20, 1917.

center in Le Havre from which he returned on February 14, 1916.

It was there that he made a request, dated February 18, 1916, to be integrated into the active army. With the support of his commanding officer, his request was accepted and he was immediately appointed second lieutenant on a permanent basis. He went back to the trenches for two months of fighting. But on April 23, 1916, he was again sent to the hospital in Compiègne to be treated for typhoid fever. His return to the front on June 1 was short-lived: he was hit 12 days later in the Quennevières farm sector by a torpedo explosion, four pieces of which injured his kidneys. Evacuated once again to a military hospital in Pierrefonds, he stayed there until July 12 to benefit from a convalescence leave that he spent in the capital and took advantage of it to get married on August 10, 1916 with Mademoiselle Madeleine Horiot at the town hall of the 11th arrondissement of Paris... But the honeymoon was undoubtedly disrupted by the state of health of the husband, He had to be hospitalized again at the military hospital n°53 of Vichy on September 8, 1916, from which he was released only two months later, on November 20, 1916, to return to the front.

In the meantime, he had been appointed (on October 15) to the rank of lieutenant on a permanent basis. His successive wounds probably made him unfit for infantry combat, as he was transferred to the air force on January 15, 1917 as a student pilot at the Avord school, where he received his basic flight training. Passing the tests on a Caudron, he obtained the military license n°5861 at the school of Tours on April 8, 1917, completed by the civil license (n°5907) which was given to him on May 18.

Arpheuil, for his part, was granted a second leave from November 24 to December 10, 1917, where he was reunited with his young wife. He returned to Paris shortly afterwards, on December 26, to attend a training course at the Plessis-Belleville training division, where he was trained on the SPAD fighter that was to equip the unit, which was now attached to Combat Group 16. He returned to the squadron on January 11, 1918, and was granted a new leave from February 20 to March 7, 1918.

SPA 151 was then assigned to the Oise region, where the German spring offensives were to take place. Clashes with the German air force became much more frequent and the first combat losses affected the squadron. For Jean Arpheuil, this was the occasion to inaugurate his record of achievement on May 27, 1918 when he attacked a patrol of 5 enemy fighters, probably hitting one of them, which was not certified. The next day he did it again and the air operations report book notes "*Lt Arpheuil attacked a biplane, which, after spinning, broke up in the air and fell into our lines near Ressons-sur-Matz.*" The victory was certified and barely three days later, on May 31, he won a second one in collaboration with Sgt Thelot by shooting down in flames an enemy two-seater in the French lines. He continued on his way and shot down another two-seater the next day, June 1st, at St-Léger-au-bois, in collaboration with Sgt Letu.

Above: SPAD XIII of escadrille N 151, during summer 1918. Note the unit insignia, a bird on the top of the fuselage.

These three victories in record time earned him a citation in the army order on June 19 (which only mentions his first victory), and then his nomination as a Knight of the Legion of Honor on July 3: "*Officer of great moral value. Wounded 3 times in the infantry and passed to the fighter aviation. Has always been for the pilots of the squadron an example of success and courage. Has led many battles, often returning with his aircraft riddled with bullets. Recently won his 2nd victory by shooting down in flames an enemy reconnaissance aircraft in our lines. 1 citation*". His 3rd victory earned him a 3rd citation in the army order, which was awarded to him on July 14, 1918. He won a 4th victory on the following July 20th by shooting down an enemy two-seater alone. Benefiting from a new well-deserved leave from July 25 to 31, he received his 4th citation in the army order on August 9.

At the top of his squadron's record, he was appointed on September 13, 1918 to take command of a new unit, SPA 170, stationed in Flanders and under the command of GC 23. He took command of his unit on September 28, but did not stay there long, until he won a 5th air victory on October 14, 1918, earning him a new citation. He contracted the flu, probably the Spanish flu that affected many soldiers, and had to be evacuated to the hospital in Dunkirk on October 27, 1918.

SPAD XIII n°2xxx, SPA 151, fall 1918.

Jean Arpheuil Victory List								
N°	Date	Unit	Plane piloted	Plane claimed	Place	Notes	German casualties	Confirmation
P1	27-May-18	SPA 151	SPAD	Fighter				
1	28-May-18	SPA 151	SPAD	Two-seater	Ressons-sur-Metz			Citation OA 19-6-1918
2	31-May-18	SPA 151	SPAD	Two-seater	Lassigny	With Sgt Thelot.		Citation LO 28-6-1918 - JORF 1-10-1918 p.8574 "2nd victory"
3	1-Jun-18	SPA 151	SPAD	Two-seater	St Léger au bois	With Avec Lt Letu.		Citation OA 14 Juil 1918 - JORF 11-9-1918 p.7984 "3rd victory"
4	20-Jul-18	SPA 151	SPAD	Two-seater	St Léger au bois			Citation OA 9-8-1918 "4th victory"
5	14-Oct-18	SPA 170	SPAD	Enemy plane				Citation OA - JORF 20-1-1919 p.772 "5th victory"

This marked the beginning of a long descent into hell in military hospitals and sanatoriums, made worse by the death of his mother. Placed on convalescent leave, he suffered a relapse in January 1919 and his hospitalization was prolonged. On February 28, 1919, he was discharged from his squadron and found himself undergoing treatment in Nice, where a military doctor diagnosed chronic post-infectious laryngitis and pleurisy contracted during service. He was sent for convalescence for 3 months, and on June 8, 1919 was sent to the auxiliary hospital n°6 of Pau, then to the hospital n°34 on July 1. A commission of three military doctors met on December 4, 1919 and concluded that he should be discharged from the army with a 100% disability. He was very ill with lung problems; his voice was constantly veiled, he had lost 22 kg. He was then demobilized and declared to retire to Pau. He survived only six months, dying on June 26, 1920 at 10 p.m. in the sanatorium of Cambo-les-Bains. Alone, if we are to believe the registrar who wrote the death certificate.

Above: Probable SPAD XIII of Jean Arpheuil in SPA 151.

Bernard Artigau

First names: Bernard.
Born: 28-08-1894 Licq-Athéry (Pyrénées-Atlantiques).
Father's profession: Farmer.
Military pilot brevet: 24-04-1917 (n°5894).
Units: SPA 15.
Army communiqué: 28-07-1918 for 10 victories.
Total score: 12 victories (1 drachen), 10 probables.
Died: 03-05-1968 in Buenos Aires (Argentina).

Above: Bernard Artigau portrait.

Bernard Artigau was born on August 28, 1894 in Licq-Athéry, a small village in the Basque country where his parents were farmers. He would only stay there for the first 18 months of his life, since his family emigrated to Argentina and settled in Buenos Aires, where he spent his childhood.

Spanish became his mother tongue and it was under the usual name of Bernardo that the young man grew up and developed a passion for aviation. In 1911 he was an apprentice mechanic and in 1914 he entered, together with his younger brother Pierre (Pedro), the carpentry school of the arsenal of Rio Santiago, with the probable goal of becoming a pilot in the nascent Argentine naval aviation.

However, events in Europe were to thwart this project. In August 1914 war was declared and Bernard Artigau, who had just celebrated his 20th birthday, was called up as a French citizen. It is not known if he received the call-up or if he deliberately refused to go; however, he did not join the 3rd Colonial Infantry Regiment where he had been called up and was classified as "good absentee" by the military administration, then "insubordinate" on February 9, 1915.

The young man returned to France voluntarily in April 1916, probably out of a spirit of adventure and a desire to become a pilot. It should be noted that he barely spoke French and discovered a country that was unknown to him and where he had no relatives... On May 26, 1916, he was drafted into the 144th Infantry Regiment as a private and was quickly forgiven for his delay in being called up, since the commander of the 18th military region issued a judgment of "refusal to inform" in his favor on June 3, 1916.

He only stayed a short time in his infantry regiment since he was transferred to the 2nd aviation group on September 21, 1916, probably to become a mechanic due to his professional skills. But he was finally directed to the flying schools on December 15, 1916 when he joined the 1st aviation group of Dijon where he followed his theoretical training. On February 3, 1917, he went to the Tours flying school where he obtained his military pilot's license (n°5894) on April 12, 1917. He then entered the Avord training school on April 14, where he was promoted to the rank of corporal on the following 28. Then, on May 19, 1917, he completed his training by joining the acrobatic school of Pau where he was instructed in the handling of fighter planes.

On June 4, 1917, he joined the Paris region training division and was quickly assigned to an operational unit, the SPA 15 escadrille, on June 20, 1917. The unit, whose symbol was the Bayard helmet, flew on SPAD VII fighters and was stationed in

Above: "Bernardo" Artigau in front of his SPAD XIII, coded "2", in Mont-l'Eveque on August 30, 1918. The plane also carries red/white vertical white bands in both sides of top wing.

the Aisne region, where, amalgamated with SPA 65, 84, 88, and 124, it formed the combat group n°13. The young corporal Bernard Artigau, whose superiors noted that he had difficulty speaking French, discovered the front and on July 6, 1917, he flew his first air patrol over Soissons and Berry-au-Bac. The young pilot was taken under the wing of a veteran, the ace Lucien Jailler, who had won 11 victories. The teaching of this veteran of the battle of Verdun was to prove invaluable for many young pilots of the SPA 15 who, following his example, were soon to win air victories.

SPA 15 migrated with GC 13 to Flanders on July 17, then settled in Champagne on August 12. It was there that "Bernardo" achieved his first successes. Promoted to sergeant on August 25, he attacked a German captive balloon on October 17, 1917, which his patrol leader, Lt. du Plessis, set on fire. His participation in this victory was recognized by a citation in the army order on October 28, 1917. From then on, his successes were to multiply. He successively shot down two enemy two-seaters on November 1 and December 23, 1917, the latter in collaboration with Sergeant Gabriel Guérin, the future ace of the SPA 15 squadron, which ended the war with 23 certified victories. On February 3, 1918, it was still with Guérin that Artigau won what was described as his third aerial victory by the citation in the army order that he received on February 28: this was an error, as his first victory against the captive balloon was not counted – an error that would persist in all his subsequent citations.

Promoted to adjudant (warrant officer) on March 31, 1918, he then participated with his SPAD in violent air battles to counter the German offensives of spring 1918. The GC 13 was grouped in General Duval's 1st Air Division, and as such was sent to all the hot spots of the front to obtain air superiority. Artigau won two air victories in April 1918, two more in May, two in June (where he was awarded the military medal), and one on July 22, 1918 against a Rumpler C that he shot down alone at Ville-en-Tardenois.

This victory was recognized by a citation in the Army Order on August 21, 1918 as his 10th victory (in fact, his 11th), and earned him the honor of having his name mentioned in the July 28 Army Communiqué. He won a last victory, his 12th, on October 28, 1918 during the last confrontations by shooting down a Fokker D VII fighter in the Fismes region, in collaboration with Lt. Gabriel Thomas of the SPA 88, an ace with 7 victories.

"Bernardo" Artigau in front of his SPAD XIII, coded "2", in Mont-l'Eveque on August 30, 1918. The plane also carries red/white vertical white bands in both sides of top wing.

Above: A bad quality picture, taken form *Jacques Mortane* magazine, showing the left side of Bernard Artigau's SPAD – possibly the same as last photos. This planes carries the feminine first name *Jeannette*.

He was then promoted to second lieutenant on a temporary basis. When the war ended, he had flown 444 hours of wartime flying, won 12 certified air victories and a large number of probable ones (23, according to his logbook). He was kept in the army and was awarded the Knight's Cross of the Legion of Honor on April 25, 1919, and was demobilized on September 24.

He stayed in France and found a job as a pilot by joining the Latécoère company in 1920, flying the Toulouse-Alicante route, where he accumulated nearly 600 hours of flight time during the year.

In 1921 he returned to Argentina where he found his brother Pierre. Together the two founded a small local airline, using aircraft from French military surplus: Four Sopwith 1A2, one two-seater SPAD, one Farman F 41, and one Caudron G3. The company operated for six years, during which time he made almost 180 night flights over Buenos Aires, and numerous day flights during which he transported doctors, journalists, politicians, policemen, photographers... 1500 flight hours were added to his logbook.

The French Air Force appointed him as a reserve lieutenant and in 1923 he was assigned to the reserve at the disposal of the Air Force Commander in French West Africa. However, Bernard Artigau was done with the army and did not serve any period in the reserves because of his distance.

In February 1929, he was recruited as assistant to the director of the Aeroposta Argentina SA, a subsidiary of the Compagnie Générale Aéropostale. He was in charge of recruiting personnel and surveying the terrain on the Patagonia line. During his professional activity he met up with his squadron mate Gabriel Thomas, who had become head of Aéroplace in Montevideo.

In the mid-1930s he was employed as a pilot by the Air France company in Buenos Aires and was placed in the army reserve at the disposal of the local air attaché. Promoted to the rank of Officer of the Legion of Honor on July 14, 1938, he was not remobilized during the Second World War, which he spent in Argentina, in his residence in Buenos Aires.

Dismissed from the Air Force because of his age on August 28, 1947, the former ace of 14-18 was honored by the French government, which awarded him the Aeronautics Medal in 1955 by the French ambassador to Argentina, then promoted to Commander of the Legion of Honor on July 18, 1961.
In 1967, at the invitation of the Association des vieilles tiges (pioneer pilots), he went to France to attend the ceremonies in honor of the 50th anniversary of the death of the ace Georges Guynemer. Also honored by the Argentine government for his pioneering role in the development of local commercial aviation, he died in Buenos Aires on May 3, 1968 and is buried in a cemetery in that city.

Bernard Artigau Victory List								
N°	Date	Unit	Plane piloted	Plane claimed	Place	Notes	German casualties	Confirmation
P1	1-Oct-17	SPA 15	SPAD	Enemy plane	O. Laon	With Lt du Plessis		Citation OA 28-10-1917 - JORF 6-1-1918 p.279
1	17-Oct-17	SPA 15	SPAD	Drachen	Siezy	With Lt du Plessis		Citation OA 30-12-1917 - JORF 22-1-1918 p.816
2	1-Nov-17	SPA 15	SPAD	Rumpler C	Chevrigny	With Sgt Calamai		Citation OA 30-12-1917 - JORF 22-1-1918 p.816
3	23-Dec-17	SPA 15	SPAD	DFW C	Berru / N.St Léonard	With Sgt Gabriel Guérin		
P2	30-Jan-18	SPA 15	SPAD	Rumpler C	Fort de la Pompelle (Reims)			Citation OA 28-2-1918 "his 3rd enemy aircraft"
4	3-Feb-18	SPA 15	SPAD	Enemy plane	Nogent l'Abbesse	With Sgt Gabriel Guérin		
P3	3-Feb-18	SPA 15	SPAD	Two seater				
P4	11-Mar-18	SPA 15	SPAD	Two seater	NE Berry-au-Bac			Citation OA 12-8-1918 - JORF 5-12-1918 p.10471 "4th and 5th adversary"
5	11-Apr-18	SPA 15	SPAD	Rumpler C	Grivesnes-Rollot			Citation OA 12-8-1918 - JORF 5-12-1918 p.10471 "4th and 5th adversary"
6	12-Apr-18	SPA 15	SPAD	Two seater	S. Grivesnes			
P5	12-Apr-18	SPA 15	SPAD					
P6	22-Apr-18	SPA 15	SPAD	DFW C				
P7	6-May-18	SPA 15	SPAD	Fokker Dr I	Montdidier			Citation OA 13-6-1918 - JORF 1-8-1918 p.6672 "6th and 7th airplane"
7	15-May-18	SPA 15	SPAD	Two seater	Thory	With Lt André Barcat		
P8	17-May-18	SPA 15	SPAD	Rumpler C	Roye			Citation OA 13-6-1918 - JORF 1-8-1918 p.6672 "6th and 7th victory"
8	27-May-18	SPA 15	SPAD	DFW C	Comin	With Adj Armand Berthelot		Citation OA et MM 25-6-1918 - JORF 4-9-1918 p.7764 "8th victoiry"
9	1-Jun-18	SPA 15	SPAD	DFW C	Villers-Cotterêts			Citation OA 21-8-1918 - JORF 11-12-1918 p.10654 "9th victory"

Bernard Artigau Victory List								
N°	Date	Unit	Plane piloted	Plane claimed	Place	Notes	German casualties	Confirmation
10	7-Jun-18	SPA 15	SPAD	DFW C	NE. Château Thierry			Citation OA 21-8-1918 - JORF 11-12-1918 p.10654 "9th victory"
P9	19-Jul-18	SPA 15	SPAD	Fokker Dr I				
11	22-Jul-18	SPA 15	SPAD	Rumpler C	Ville-en-Tardenois			Citation OA 21-8-1918 - JORF 11-12-1918 p.10654 "10th victory"
P10	1-Oct-18	SPA 15	SPAD	Enemy plane				
12	28-Oct-18	SPA 15	SPAD	Fokker D.VII	Fismes	With Lt Gabriel Thomas (SPA 88)		Citation OA et LH 25-4-1919 - JORF 8-3-1919 p.2496 "11th victory"

SPAD XIII (serial unknown) "2", SPA 15, Mont-L'Evèque, May 30, 1918.

Albert Auger

First names: Albert, Victor, Robert.
Born: 26-01-1889 in Constantine (Algeria).
Father's profession: Army officer (captain).
Military pilot brevet: 11-05-1915 (n°928).
Units: C 11, N 31, N 3.
Army communiqué: 17-05-1917 for 5 victories.
Total score: 7 victories, 9 probables.
Died: 28-07-1917 in Dunkirk (Pas de Calais), died from his wounds in hospital after aerial combat.

Above: Albert Auger portrait.

Son of a senior officer, destined to serve in the army

Albert, Victor, Robert Auger was born in Constantine, Algeria, on January 26, 1889, into a very military environment: his father was Captain Michel Armand Auger, then 34 years old, captain in the 3rd Zouave Regiment. The son of a surveyor from the Eure region, Captain Michel Auger enlisted at the age of 19 and was appointed an officer after attending St-Cyr. He distinguished himself during a one-year assignment in Africa and then four years in Algeria where his son Albert was born. Upon his return to France in 1891, he was awarded the Knight's Cross, then the Officer's Cross of the Legion of Honor, and was given various assignments, including in 1904 the position of Chief of the Technical Section of the Infantry, earning his Brigadier General's star in 1910 and being appointed Deputy Chief of the General Staff of the Army in April 1911.

The young boy grew up with his 3 brothers and sisters according to his father's transfers, and at the age of 8 he left the family home to study at the Lycée Buffon in Paris as a day student, where his father had obtained a scholarship. We know from his early years that he had a passion for sports and showed a talent for drawing and watercolor, winning a prize in 1901 for one of his works exhibited at the Petit Palais. Following a path laid out by his father, the young man of 1m64 enlisted in the army at the age of 18 by presenting himself at the Town Hall of Montauban on October 24, 1907 where he signed a 3-year contract and chose to be incorporated into the 11th infantry regiment encased in this city.

He spent the first two years of his military life there, rising rapidly in rank as he was promoted to corporal on January 24, 1908, to sergeant on the following September 26 and to sergeant-fourrier on April 14, 1909. On October 26, 1909, he was transferred to the 8th Battalion of Foot Chasseurs, an elite infantry unit, stationed in Amiens. He re-enlisted there at the end of his contract on November 1, 1911 and prepared for the competitive examination for the infantry military school. On June 14, 1911, the official journal announced his admission to the oral exams, which he passed. He entered the school on November 1, 1911 as a midshipman and left a year later with the stripes of second lieutenant, being assigned to the 3rd company of the 31st infantry regiment in Paris, which he

Right: Many SPA 3 aces on this photo: Extreme left, Georges Raymond (5 victories). Then, on the foreground, from left to right, are René Dorme (23 victories), Albert Deullin (20 victories) and Albert Auger.

joined on October 1, 1912.

Wounded during the first battles, he was transferred to the air force

He was still there when the war broke out. When the war broke out, the 31st Infantry Regiment was sent to the front of the French Third Army in the Ardennes, as part of the 10th Infantry Division commanded by General Michel Auger. The French troops were going to lead disastrous infantry charges, without any artillery preparation, against German troops defending themselves from the fire of their machine guns. The 31st Infantry Regiment had its baptism of fire at Cutry-Rehon on August 22, 1914, and suffered serious losses to the point that Second Lieutenant Auger was put in charge of his company because of the wounds suffered by the other officers. On August 31, 1914, the regiment was engaged south of Vaux en Dieulet in what would become known as the Battle of Fossé, and came under heavy musketry, machine-gun, and artillery fire. The young officer, caught under enemy fire, was hit by three bullets in the legs, one of which broke his ankle. Evacuated, the doctors diagnosed a "*sethon wound of the right thigh with probable incomplete fracture of the femur, not infected.* "

Sent to convalesce, he was promoted to the rank of lieutenant on October 1, 1914 and was awarded the Knight's

Left: Albert Auger in a SPAD VII. The photo is signed by SPA 3 top aces : Heurtaux, de la Tour, Guynemer, and Deullin.

Cross of the Legion of Honor on November 8. But despite his recovery, he never regained his physical shape and proved unfit to return to combat in the infantry.

Burning with the desire to return to combat, he volunteered for the air force, which was recruiting volunteers in all arms at that time. He became a student pilot on January 25, 1915, and took flying lessons at the Pau school where he obtained his civilian pilot's license (n°1918) on May 6, 1915 on a Blériot monoplane, followed by his military pilot's license (n°928) on May 11.

Fighter pilot in the N 31 squadron

He was affected to C 11, a reconnaissance unit, but did not stayed here long. While the fighter aviation was being formed and the first fast Nieuport 10 aircraft were delivered to some units, Lieutenant Auger left the C 11 squadron on July 29, 1915 to train to fly twin-engine Caudron G4 aircraft, but did not return to his unit. Discovering the Nieuport 10 in the stocks of the general aviation reserve, he obtained on August 10 to convert to this new aircraft and was transferred to the N 31 squadron, stationed at Toul, which he joined on August 9, 1915.

The squadron, led by Captain Louis Mathieu, was equipped with two-seat Nieuport 10s that replaced the last Morane Parasols. Albert Auger took command of the squadron on September 22, 1915, under tragic circumstances: that day, he took off on a two-seat Nieuport with Lt Gustave Lagache as observer, and accompanied another Nieuport flown by his squadron leader and taking Captain Gabriel Petitjean as observer. The two aircraft had the mission of attacking German Drachen, but encountered four German aircraft during their mission. The French fought one against two, and, despite the interposition of Auger's aircraft, Captain Mathieu's Nieuport was shot down.

Returning to his base on his Nieuport, which had probably been hit by enemy bullets, Alfred Auger, the oldest active officer with the highest rank, took over command of N 31. Rewarded with a citation in the army order (his 3rd) for the fight of September 22nd and promoted to the rank of captain on December 28th 1915, his squadron received its first single-seat Nieuport 11s specifically designed for hunting in early 1916.

Although still based in Lorraine at Manoncourt en Vermois, N 31 maintained a detachment at Ancemont in the Verdun region, where aerial reconnaissance, notably by the C 11 squadron, showed significant German reinforcements converging on the area. A gigantic German attack was indeed launched on February 21, 1916, and the few fighters of N 31 battled the German air force from day one. Auger recounts: "*I found myself with some comrades being attacked by four Boche. One of them was holding me close, and I came down on his face. I missed him. I do not shoot. I meet another one in the same conditions and after a few turns I fall by chance 10 meters under an L. V. G. who had not seen me. I fired, he went down, but was able to land. He was shot down, but not in pieces.* "This victory would not be recognized, but he would soon have the opportunity to find new prey, as the aerial battles over Verdun took on an importance never before known.

On March 13, 1916, he shot down an LVG C at Cumières: "*Looking for a Boche in the Verdun region at 4,000 meters, I saw one passing through the lines around 2,000 meters. Immediately, I went down hard on him, and, when I arrived at 10 meters, I fired right in the middle. The observer was killed on the spot, his machine gun went up to the sky. The pilot was killed on his controls and fell forward. The LVG executed a superb rollover, then fell on his back. At that point one of the two Germans was thrown about 1500 meters and I came back singing a little tune....* "According to the German archives, the aircraft could be a two-seater of the KG 2 piloted by Lt.d.R. Wilhelm Otto, killed that day at Douaumont, 15 km from Cumières. The victory was recognized and rewarded by a 4th citation in the Army Order, published in the official journal of April 21, 1916.

On April 2, 1916, escorting a C 11 aircraft piloted by Captain Vuillemin, he won his second certified victory, as he described in a letter: "*On April 2, I left with Captain Vuillemin, my former squadron leader. All of a sudden, I saw an Aviatik coming down on him at a steep angle. I turned back to the Kraut. At point blank range I fired. The Aviatik went down immediately and crashed to the ground. The captain gave me signs of intense satisfaction.*" The Aviatik, shot down near Récourt-le-Creux, was confirmed. The next day, April 3, he was confronted with a Fokker Eindecker fighter: "*I left with Captain Vuillemin. Suddenly, I saw a superb Fokker that threw itself at him. I turn around again and I fall on the Fokker at 10 meters. I aim at the pilot between the two shoulders. I barely had time to shoot because I was going at a crazy speed. I missed catching the wheels of the Fokker with my wing. I fell to the right and he fell to the left, only he had bullets in his skin. I straightened up, but he continued to the ground. He was shot down, but not in pieces.*" Landing near St-Maurice-sous-les-Côtes, the Fokker was not certified, but the German archives inform us that Oberleutnant Bruno Loerzer, a Fokker E pilot at KEK Jametz on the Verdun front, was seriously wounded in aerial combat. He was later awarded a 5th citation in the army order for these two battles.

But his luck ran out on April 16, 1916 when his Nieuport crashed under unknown circumstances. Alfred Auger survived the crash with a broken jaw and a severe concussion. Evacuated to the rear, he was officially removed from his squadron on April 28 and began a new long convalescence...

He joined the Cigognes squadron

He had to fight to return to the front: while he was recuperating, he already contacted Captain Félix Brocard, commander of the elite N3 squadron and head of the Cachy Fighter Group, which was fighting on the Somme front. Barely recovered, he joined the Groupement des Divisions d'Entrainement on September 10 to retrain as a pilot, and left on October 15, 1916 "to the 6th Army at Cachy", where officially he was not integrated into any squadron but unofficially flew on a Nieuport 17 (n°1499) of the N 3 squadron. He is described there at that time by a witness as "*a small, skinny, blond man, still poorly recovered from a recent wound, but he came to spend his convalescence leave at his squadron, and as he would be too bored there if he didn't fly, he hunts there like the others,*" or again "*Original, strange, very cultured, gifted with extraordinary courage, inseparable from Commander Brocard.* "

Brocard held him in very high esteem and entrusted him with one of the rare SPAD VIIs of the squadron (n°158) on which he exercised his artistic talents by painting himself a personalized stork, which held a miniature German soldier in its beak. Making his first flight on board on November 9,

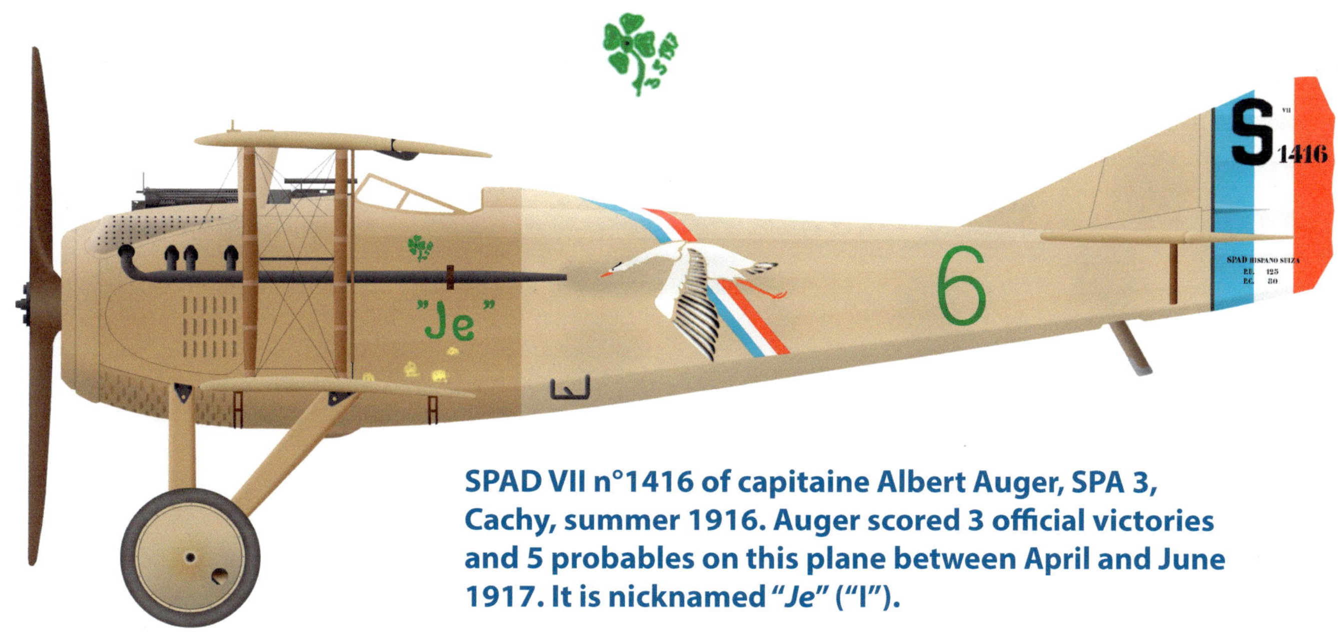

SPAD VII n°1416 of capitaine Albert Auger, SPA 3, Cachy, summer 1916. Auger scored 3 official victories and 5 probables on this plane between April and June 1917. It is nicknamed "*Je*" ("I").

1916, he was officially integrated into the N 3 squadron, now commanded by Captain Heurtaux, but, victim of frequent jamming, he did not win any victories. During the night of November 16th to 17th, his aircraft was destroyed on the ground by a bombing raid and he remained without flying for some time, before being assigned a new aircraft, the SPAD VII n°170, probably the aircraft on which he won an unregistered victory on January 25, 1917 against a two-seater in the Peronne area, before the N 3 squadron left the Somme for the Manoncourt-en-Vermois airfield in Lorraine, a sector he knew well because he had fought there a year earlier with the N 31.

Captain Auger got into the habit of patrolling alone and on February 9, 1917, near Rogéville, he won his third confirmed victory under circumstances described by a witness, Jacques Boulenger: "*Today, as he was going around the lines to amuse himself, not on his SPAD which was being repaired, but on an old Nieuport that he had borrowed, he met the German patrol that was coming back after having thrown a few bombs here and there. He was "a little annoyed" because his machine gun, out of tune, could only fire one shot at a time, but the Germans did not see him; they were not very suspicious, being in force, so he started to follow them. At a certain point, one of them moved aside; immediately, he poked at him and "photographed" him. And that's how Führer Krone went down in flames. It's very simple.*"

Rewarded with a sixth citation in the army order, his habit of attacking alone almost cost him his life on the following February 16 when he attacked 4 single-seaters over Lunéville, which gave him a severe beating and forced him to land in the wilderness with "bullets all over the place - slight scratches". What he does not mention in the squadron's logbook is that a bullet went through his wallet and was only stopped by the wad of bills that was in it...

Flying on the SPAD VII n°225 recovered from the adjutant Chaînat who had been dismissed from the squadron, he also recovered the n°6 which he would repeat on his other aircraft and especially the n°1416, a new aircraft with a 200 hp over-compressed engine which he received on April 14, 1917 and with which he would obtain most of his successes, while the N 3 squadron was set up at Bonnemaison in anticipation of the attack on the Chemin des Dames.

It was precisely at the beginning of this battle, on April 16, 1917, that he shot down a two-seater on Amifontaine that he noted "may have been forced to land" and which was not certified, nor was the fighter he shot down during a patrol with the ace René Dorme, attacked among a group of 4 single-seaters.

On April 22, he won his fourth official victory by shooting down a two-seater over Lierval, which, according to the German archives, might well have been an FA 212 aircraft whose crew, Uffz. Gustav Richer and Lt. Erich Bersu, had fallen at Chevregny, near Lierval. Rewarded with a new citation in the army order (his 7th), he won a new probable victory on May 2, 1917 against a plane that he had to let go, but from which he received a bullet that went through the cockpit and just missed him... His SPAD VII n°1416 will now be decorated with a flower painted on the impact site!

Ace of war and commander of the N 3

On May 5, 1917, Captain Heurtaux, commander of the N 3 squadron, was wounded in combat and had to be evacuated. The next day, Auger officially took over from him, thus becoming the leader of the most famous squadron in France.

A few days later, he was to become the equal of the prestigious pilots who made up the squadron: on May 11, 1917, he won his 5th confirmed victory against a two-seater shot down at Vailly in collaboration with Captain de Sevin, of N 12. He was now an ace and as such was distinguished by seeing his name mentioned in the following May 17 communiqué to the army. This earned him a little fame, since the newspapers of the time took up the communiqué and published his biography on their front pages, hailing the arrival of the new ace.

Albert Auger was to win further successes in the Aisne, still aboard his SPAD VII n°1416: a probable victory on May 24th, a 6th official success on June 4th 1917 against an aircraft shot down at Grandelain, another probable success on June 17th, and then a 7th official success against an aircraft shot down on June 28th 1917 to the west of Pontavert

On July 11, 1917, the n°12 Combat Group, which was part of the N 3 squadron, left the Aisne front for Flanders where it was to reinforce the British Air Force and the British Army conducting an offensive in the area. Settling on the Bergues airfield, the French fighter pilots faced a German air force at the peak of its efficiency, made up of experienced fighters having won impressive fighter charts against technically inferior British aircraft, and using proven fighter tactics by going out in patrols of six aircraft.

The French fighter pilots belatedly became aware of this fact and organized themselves into comparable patrols. Auger, an ace and squadron leader, did not give up solo patrols, which was fatal for him on July 28, 1917. Attacking alone a group of five enemy fighters, he was hit in the air by a bullet that hit him in the ear and went out through his nose. Losing a lot of blood, he managed to get out of the fight and land in the Allied lines between Woesten and Zuydchotte where he died shortly after. His victor could have been Vfw. Rudolf Francke who claimed a victory that day that was not certified to him at Boesinghe, near Woesten. He ended the war with 14 official victories.

Captain Auger's funeral was organized on August 2, 1917 in the chapel of the Dunkirk hospital, in the presence of his father, military dignitaries, delegations from various squadrons and the entire N 3. Major Brocard, head of GC 12, gave a poignant speech. Albert Auger, who died in combat at the age of 28, and who had won 7 victories, was only the first of the great French aces to fall in Flanders: Deullin and Heurtaux were wounded shortly after him, while the ace of aces Georges Guynemer, followed him into the grave on 11 September 1917.

Albert Auger Victory List								
N°	Date	Unit	Plane piloted	Plane claimed	Place	Notes	German casualties	Confirmation
P1	21-Feb-16	N 31		Enemy plane	Chaussée			
1	13-Mar-16	N 31		LVG C	Cumieres		KG 2 - Ltn Wilhelm Otto KIA Douaumont	Citation OA - JORF 21-5-1916 p.4464
2	2-Apr-16	N 31		Albatros D	Creux			
P2	3-Apr-16	N 31		Fokker E	St-Maurice		Oblтn Bruno Loerzer (KEK Jametz) WIA	
P3	25-Jan-17	N 3	?	Two-seater	Péronne			
3	9-Feb-17	N 3	Nieuport	Albatros	Rogeville			Citation OA - JORF du 18-7-1917 p.5558 "3rd airplane shot down"
P4	16-Apr-17	N 3	SPAD VII n°1414	Two-seater	Prouvais / Amifontaine			
P5	19-Apr-17	N 3	SPAD VII n°1416	Fighter	Orangeville			
4	22-Apr-17	N 3	SPAD VII n°1416	Two-seater	Lierval		FA 212 - Uffz Gustav Richer et Ltn Erich Bersu, KIA Chevregny near Lierval	Citation OA - JORF du 9-8-1917 "4th enemy airplane"
P6	2-May-17	N 3	SPAD VII n°1416	Enemy plane		Wounded by 3 bullets		
5	11-May-17	N 3	SPAD VII n°1416	Two-seater	Vailly	With Lt Joseph de Sevin		Citation OA - JORF 11-5-1917 p.7344 "5th enemy airplane"
P7	24-May-17	N 3	SPAD VII n°1416	Two-seater	Fresnes			
P8	25-May-17	N 3	SPAD VII n°1416	Two-seater	Berrie			
6	4-Jun-17	N 3	SPAD VII n°1416	Enemy plane	Grandelain			SPA 3 logbook - confirmed
P9	17-Jun-17	N 3	SPAD VII n°1416	Enemy plane				
7	28-Jun-17	N 3	SPAD VII n°1429	Enemy plane	O. Pontavert			SPA 3 logbook - confirmed

Paul Barbreau

First names: Paul, Augustin, Edouard.
Born: Renault (Today Sidi M'Hamed Ben Ali–Algeria).
Father's profession: Colonial civil servant.
Military pilot brevet: 14-05-1917 (n°6406).
Units: N 154.
Army communiqué: No.
Total score: 8 victories (8 Drachens).
Died: 02-06-1976 in Marseilles (Bouches-du-Rhône).

Above: Paul Barbreau with his commanding officer on the right, the great ace Michel Coiffard (34 victories, one of the top balloon busters).

Paul Augustin Edouard Barbreau was born on September 16, 1894 in the town of Renault in French Algeria (in the department of Oran), today called M'Hamed Ben Ali. He came from a family of colonial civil servants whose father Pierre was an assistant administrator and who moved to Algiers at an unknown date. The child, who reached the height of 1m71 in adolescence, receives a good level of education and very probably obtained his baccalaureate, which allowed him to undertake law studies.

He was still a student when the war broke out and was drafted into the army with his age group on September 16, 1914, a month and a half after the fighting began in France. He was drafted as a private 2nd class into the 5th regiment of chasseurs d'Afrique and was to experience the trenches of the front. The cavalry regiments being rendered useless by the war of position, they sent detachments to stand guard in the trenches. Paul Barbreau, whose level of education quickly allowed him to reach the rank of corporal on November 18, 1914, and then marshal of logis on February 6, 1915, spent two and a half years in the trenches during which he served at Verdun and on the Somme. During this period, he left his cavalry regiment to be assigned to a marching battalion of African light infantry. It was there that he was distinguished by a citation that earned him the Croix de guerre with a bronze star on May 14, 1916: "*Animated by a great sense of duty, he remained for 8 days in an observatory that was continuously shelled, and ensured the observation despite the displacement of the battlements by a shell that hit it head-on.*"

After volunteering for aviation, he was affected on April 1, 1917 as a student pilot in Dijon and then to the Ambérieu flying schools. It was there that he passed his military pilot's license, obtaining excellent marks in the theoretical exam (nearly 18/20 in all subjects) and then the license n°6406 on May 14, 1917. He was then assigned to the Avord school on June 22, to the Pau acrobatics school on August 14, and did a shooting course at the Cazaux school on September 1, 1917.

He was then assigned to the Groupement des Divisions d'Entrainement on September 15, 1917, awaiting his assignment to a squadron, which came two weeks later when he joined the N 154 squadron on October 1, 1917, where he carried out his first missions on the front without any particular distinction.

However, at the beginning of 1918, a new squadron leader, Lt. Lahoulle, specialized in hunting German observation balloons, the Drachen, a vulnerable target in itself but very dangerous to approach because it was strongly defended by the flak, not to mention protective aircraft. After having set the example by personally shooting down a first Drachen on April 1st 1918, Lahoulle was followed by several of his pilots, particularly S/Lt Coiffard, who saw it as a springboard to the glory he was seeking and indeed obtained by winning 34 aerial victories before falling in battle shortly before the armistice.

Above: Paul Barbreau is 2nd from the left among those SPA 154 pilots standing behind a SPAD XIII. The aces Ehrlich (19victories) is the 7th, just under the propeller.

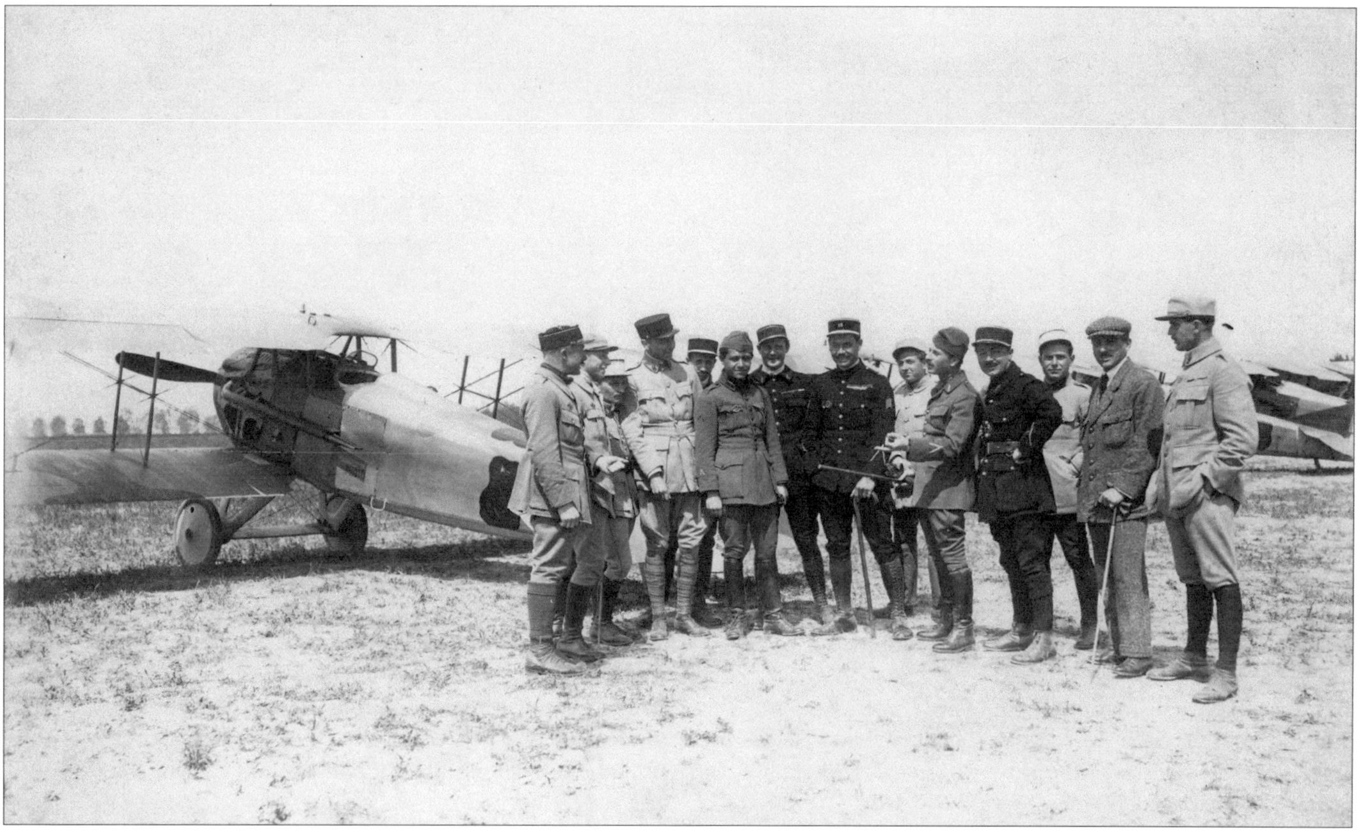

Above: SPA 154 pilots and many aces among them: Paul Barbreau (6th from left), Michel Coiffard (34 victoires, 8th) and Jacques Ehrlich (19 victories, 10th).

Paul Barbreau Victory List								
N°	Date	Unit	Plane piloted	Plane claimed	Place	Notes	German casualties	Confirmation
1	2-Jun-18	SPA 154	SPAD VII n°3214	Drachen	Igny l'Abbesse	With S/Lt Waddington (3)		
2	5-Jun-18	SPA 154	SPAD XIII n°8482	Drachen	Trigny	With Cne Lahoulle (7), S/Lt Waddington (4) and S/Lt Gros (3)		
3	30-Jun-18	SPA 154	SPAD XIII n°8482	Drachen	Beauvardes	With MdL Ehrlich (1) and S/Lt Coiffard (4)		Citation OA - JORF 3-9-1918 p.7725 "3rd victory"
4	17-Jul-18	SPA 154	SPAD XIII n°8482	Drachen	Beine	With S/Lt Waddington (6)		
5	18-Jul-18	SPA 154	SPAD XIII n°8482	Drachen	Forêt de Ris	With Sgt Abott (2)		Citation OA - JORF 23-10-1918 p.9199 "Shot down 5 enemy devices"
6	1-Aug-18	SPA 154	SPAD XIII n°8482	Drachen	N. Sommepy	With S/Lt Waddington (8) and Adj Chevalier (3)		
7	3-Aug-18	SPA 154	SPAD XIII n°8482	Drachen	N. Sommepy	With S/Lt Coiffard (21)		
8	3-Aug-18	SPA 154	SPAD XIII n°8228	Drachen	N. Sommepy	With S/Lt Coiffard (22)		Citation OA - JORF 29-12-1918 p.11227 "put to 8 the number of his victories by shooting down 3 Drachens"

At the beginning of the summer of 1918, the entire squadron was to specialize in attacking the Drachen, which the press described as being "like dry straw in the sun" under the fire of the pilots of the 154.

Paul Barbreau personally contributed by burning 8 Drachen, all in collaboration with other pilots, in a short period of two months from June 2nd to August 3rd 1918. Promoted to second lieutenant on a temporary basis on July 30, 1918, he was rewarded with a long leave to visit his parents in Algeria, during which he was awarded the military medal on August 9, 1918.

On his return to France on September 17, 1918, he was transferred to the SPA 164 squadron, a new unit, but left it on September 29 to return to the SPA 154 squadron where he finished the war and of which he was one of the few aces to survive. He remained in his squadron after the armistice and obtained a new leave for Algeria from December 4, 1918 to February 2, 1919, during which he was definitively confirmed in his rank of second lieutenant. Assigned to a flying school when the squadron was disbanded on March 16, 1919, he was demobilized shortly afterwards and returned to civilian life, being decorated with the Legion of Honor on July 7, 1919.

Posted to the reserve in Algeria, he moved to the south-east of France after some time. On April 30, 1921, he got married at the town hall of Le Muy in the Var and started a family from which three boys were born. He obtained a job as managing director of the Vassas Frères et Cie company in Marseille and settled permanently in that city. The army, which had promoted him to the rank of lieutenant, removed him from the Algerian reserves in 1924 and found him other positions in metropolitan France. However, he did not undergo any training period and quickly lost his piloting skills to such an extent that he was classified as a class C pilot, i.e. limited to the piloting of small liaison aircraft, in the early 1930s.

A Drachen, Paul Barbreau's favorite target.

When the Second World War broke out, he was mobilized on September 2, 1939 with the rank of lieutenant, then assigned on December 1 to the Air Battalion 124, at the Cazaux base, where he served as an air defense instructor. On February 28, 1940, he was removed from the air force for being physically unfit and was appointed captain on March 15, 1940. Evacuating the Cazaux base after the armistice, he was demobilized on July 20, 1940 and retired to his home in Marseille, returning to his private affairs for good, and according to his military record, taking no part in resistance activities.

Promoted to the rank of Officer of the Legion of Honor on October 31, 1961, he died in Marseille on June 2, 1976.

André Barcat

First names: André, Jean, Louis.
Born: 08-11-1896 in Paris (17).
Father's profession: Mason.
Military pilot brevet: 26-11-1917 (n°10 069).
Units: Spa 155, Spa 153.
Army communiqué: No.
Total score: 6 victories (1 drachen), 2 probables.
Died: 16-07-1918 (Marne) (Killed in aerial combat).

Above: Portrait of André Barcat.

André Jean Louis Barcat was born on November 8, 1896 in the 17th arrondissement of Paris, into a working-class family: his father was a small masonry contractor and the witnesses who ratified his birth certificate were one of his uncles, a mason, and a family friend who worked as an accountant.

The young boy who grew up in Paris was to prove to be very gifted for studies, since he obtained his baccalaureate and prepared to enter the Central School of Arts and Manufactures with the aim of becoming an engineer. But the war came and the young man who was not yet 19 years old was mobilized on April 12, 1915 in the 30th artillery regiment. However, he was registered for the entrance exam to Centrale engineer school and was able to take the exams which took place from May 7 to 8, 1915, and passed in the middle of the ranking. Because of his success in the competitive examination at the Ecole Centrale, he was immediately sent to a platoon of cadets at the Artillery Training School in Fontainebleau. He left the school on October 1, 1915 with the rank of *aspirant*, and was sent to the 36th field artillery regiment, but was also immediately assigned to an air observer course. This was carried out at the Dugny-le-Bourget center and ended on November 3, 1915, when he was sent back to the artillery depot to await his assignment to a unit.

Still attached to the 36th artillery regiment, he served during 1916 in several observation squadrons to carry out artillery adjustment missions: he is mentioned in F 62, V 21, F 19, and F 206 according to his record in the general aeronautical reserve, but without being found in the accounting books of these units. After this year of "training", and after having followed additional courses at the artillery school of Fontainebleau, he was appointed on September 27, 1916 to the rank of second lieutenant on a temporary basis, and from this date on, he was transferred to the 83rd artillery regiment.

Returning to his artillery observation missions, he volunteered to join a pilot school, which was accepted and effective on October 13, 1917, when he joined the Chartres school. It was there that he obtained his military pilot's license (n°10 069) on November 26, 1917. He proved to be quite gifted at piloting and, after passing through the Avord

Right: André Barcat standing in front of his SPAD VII.

advanced school, he went to the Pau acrobatics school to familiarize himself with the handling of fighter planes. He then moved to the training division group in the Paris region, awaiting his assignment to an operational unit.

He then received his assignment order to the SPA 155 squadron, which he joined on February 11, 1918, but he stayed there for only 11 days before being transferred to the SPA 153 squadron stationed at the same Villeneuve-les-Vertus airfield. The two squadrons, associated with the SPA 48 and SPA 94, formed the Groupe de Combat n°18 commanded by Captain Sabattier de Vignolles. The GC 18 was one of the components of General Duval's air division, a great air armada in charge of ensuring air superiority on any point of the front chosen by the general staff.

This air force was very quickly used to counter the air offensives launched by the Germans in the spring of 1918, hoping to win the war before the arrival of the American troops, and committing their last reserves. On the ground as well as in the air, the fight was going to be fierce and the French fighter pilots had their work cut out for them.

André Barcat, who went on leave from February 26 to March 1, was given his first aircraft on March 9, 1918, the SPAD VII n°3127, which he took over from Sgt Emile Jannekyn, who was transferred to another unit. He flew his

Left: André Barcat in the cockpit of his SPAD VII.

Below: Pilots of GC 18 (joining escadrilles SPA 48, SPA 94, SPA 153, and SPA 155) rest on their airfield's grass. André Barcat is 5th from the left. Two other aces: 7th from the left, in the background, is Georges Hallberger (5 victories, SPA 153). The 8th is Gilbert Deguingand (9 victories, SPA 48). On the background are SPAD VII and XIII of escadrille SPA 48, decorated with their insignia, an angry rooster in a circle.

André Barcat Victory List								
N°	Date	Unit	Plane piloted	Plane claimed	Place	Notes	German casualties	Confirmation
1	21-Mar-18	SPA 153	SPAD VII n°3127	Albatros D	Mont Cornillet	With S/ Lt Gilbert Deguingand (SPA 48)		
2	2-May-18	SPA 153	SPAD VII n°3017	Pfalz D III	Boissière, near Montdidier	With Lt Auguste Lahoulle (SPA 154), S/Lt Ernest Louvat, Sgt Rémy Morel.	Jasta 76, Ltn Moritz Weig POW.	Citation OA - JORF 1-8-1918 p.6672 "2nd enemy plane"
P1	2-May-18	SPA 153	SPAD VII n°3017	Pfalz D III	Montdidier	With S/Lt Ernest Louvat.	Jasta 79b, Ltn Fritz Edler von Braun WIA	
3	15-May-18	SPA 153	SPAD VII n°3017	Rumpler C	N. Thory	With Adj Bernard Artigau (SPA 15)		
4	19-May-18	SPA 153	SPAD VII n°3017	Chasseur	Ailly (S. Moreuil)	With MdL Arrault.	Jasta 46, Ltn Aloïs Weber POW.	
5	2-Jun-18	SPA 153	SPAD VII n°3017	Drachen	E. Thory	With MdL Georges Halberger		Citation OA - JORF 20-8-1918 p.7339
P2	14-Jun-18	SPA 153	SPAD VII n°3017	Fokker D VII	St-Pierre-Aigle			
Citation LO - JORF 16 Jan 1920 "Killed in action on July 15, 1918, after having shot down his 5th victory".								

first patrols with it and quickly acclimatized to the front, to the point where he won his first aerial victory barely two weeks later, on March 21, 1918, when he shot down an Albatros D V fighter on Mont Cornillet in the company of an ace of the SPA 48, S/Lt Gilbert Deguingand.

He then received a new aircraft, the SPAD VII n°3017, and distinguished himself on board during the month of May in the Somme, where the front was stabilized after the second German spring offensive had pulverized the British lines. On May 2, 1918, he shot down two Pfalz D IIIs in collaboration, one of which was awarded a citation in the army order because the German pilot, who had landed in the French lines, was captured. Then on May 15, a Rumpler C fell under his combined fire with that of SPA 15 ace Bernard Artigau. On May 19, 1918, he shot down a new fighter with a teammate, who was captured, and then, on June 2, 1918, he ignited a Drachen, which was certified as his 5th victory and earned him a new citation in the army order. On June 14, 1918, he closed his hunting record on his SPAD VII n°3017 by shooting down a Fokker D VII, but which was not confirmed.

After this date, he left for a well-deserved 10-day leave on June 17, 1918, returning to his unit on July 1. During his time off, he was confirmed permanently in his rank of second lieutenant on the 24th. On his return, he flew on a SPAD XIII, n°4444. But this aircraft did not bring him any luck: on July 16, 1918, while the GC 18 was fighting in the skies over Champagne, he came across a patrol of several Fokker D VIIs of Jasta 19, which attacked the French formation and shot down two SPADs near Malmy. Second Lieutenants André Barcat and Georges Lutzius did not return.

The body of André Barcat was found by the French troops and was buried in the Revigny necropolis, before being moved to a family vault. André Barcat died for France at the age of 21 and was posthumously awarded the Legion of Honor on January 16, 1920.

Bernard Barny de Romanet

First names: Bernard, Henri, Marie, Léonard.
Born: 28-01-1894 in Saint-Maurice-de-Satonney (Saône-et-Loire).
Father's profession: Land owner.
Military pilot brevet: 04-01-1916 (n°2248).
Units: C 51, N 37, SPA 167.
Army communiqué: 25-08-1918 for 10 victories.
Total score: 18 victories (1 Drachen), 3 probables.
Died: 23-09-1921 (Marne) in Etampes (Essonnes) (killed in a flying accident).

Above: Bernard Barny de Romanet, taken postwar in front of his SPAD XXbis racer in the Buc meeting, October 1920.

An aristocrat's son

Bernard, Henri, Marie, Léonard Barny de Romanet was born on January 28, 1894 in the château of St Maurice de Satonnay, a Burgundian village surrounded by vineyards. His father Léonard was a former cavalry captain who had become a gentleman farmer in the small château he had bought two years earlier and where he maintained his stable of racehorses, assisted by his servants (the château had up to nine). His mother, born Isabelle de Veyssière, gave birth to three other children of whom Bernard was the second: Marie-Louise (born in 1886), Pierre (born in 1896) and Chantal (born in 1898).

The young man, who was taught horseback riding by his father, did well at the local school and was sent to the Collège des Minimes in Chalon-sur-Saône, then to the Collège de Mongré in Villefranche-sur-Saône, run by the Jesuits. After obtaining his first baccalaureate, he passed his second (mathematics) at the Lamartine high school in Mâcon.

Career rider

Once he passed his 18th birthday, he decided to follow his father's path and went to the Mâcon City Hall to sign up for a 5-year commitment in the army on December 6, 1912. He was immediately incorporated into the 16th regiment of horsemen stationed in Beaune. Very well noted by his hierarchy, he quickly gained rank by being appointed brigadier (cavalry corporal) on June 19, 1913, then *maréchal des logis* (cavalry sergeant)on November 25.

When the war broke out on August 2, 1914, the 16th Regiment of Mounted Chasseurs was engaged on the Vosges front by half regiments. Bernard Barny de Romanet left with the 2nd half-regiment on August 16, 1914 and found himself in the Vosges, where his unit was quickly overrun by the enemy advance. He distinguished himself on September 8, 1914 and received a citation in the order of his regiment for having, in the village of Mattexey (south of Lunéville), "*while his platoon was exposed to heavy fire from the enemy infantry, did not hesitate to dismount to relieve one of his wounded cavalrymen to allow him to join his unit.*"

The 16th Regiment of Mounted Chasseurs was then

Above: General Gouraud shakes hand of lieutenant Bernard Barny de Romanet in 1918.

directed to the Meuse at St Mihiel on September 15, 1914, before being sent from October 25 to December 2 to the Yser in the race to the sea. The trench warfare made the cavalry regiments useless and the unit, back in St Mihiel from January 1915, had to provide a detachment of its cavalrymen to occupy the trenches as mere foot soldiers. It was there that *maréchal des logis* Barny de Romanet had the pain of learning of the death of his father, who died at the Hôpital-Dieu in Beaune of bronchitis contracted at the front where he had been mobilized. Probably tired of the uselessness of the cavalry in a modern war, he asked to join the air force and was accepted as a student pilot on September 20, 1915.

Passage to the air force

He went directly to the Pau aviation school on October 4, 1915 and passed his military pilot's license (n°2248) on January 4, 1916 on a Caudron aircraft. He was then sent to the Paris region to the General Aviation Reserve (RGA) on January 19, then to the Training Division Group (GDE) on February 7. His first operational assignment came on February 7, 1916, to the C 51 squadron, an observation unit attached to the 1st colonial army corps equipped with Caudron G3 and G4 aircraft.

Bernard Barny de Romanet joined the unit on April 24, 1916 and fought there for almost nine months, during which he was granted a short leave on August 27, three days after being promoted to *adjudant* (warrant officer), and another one from November 17 to December 1, 1916. He distinguished himself by carrying out numerous photographic missions while taking fire from enemy flak and hunting, which earned him a citation in the army order on October 17, 1916, and one in the army corps order on January 28, 1917.

The latter citation mentions that he was designated to fly a fighter plane: he actually left his unit the day after tomorrow to return to the GDE and learn to fly Nieuport fighters.

Fighter pilot

He was assigned to the N 37 fighter squadron based on the Champagne front, which he joined on February 9, 1917. The unit was a component of the GC 15 and took part in the battle of the Chemin des Dames in April 1917. It was during this battle that Bernard Barny de Romanet won his first victory on May 3, 1917, by attacking two enemy aircraft in the German lines and shooting down one of them on the Craonne plateau.

This exploit earned him the Military Medal on May 23, and, after a leave of absence from June 5 to 15, he was promoted to second lieutenant on July 6, 1917. His squadron, entirely equipped with SPAD fighters, was then engaged at Verdun for the recapture operations launched at the end of 1917, but did not meet with much success – de Romanet claimed two aerial victories, but these were not certified.

Its score would only really increase during the German offensives of spring 1918. The SPA 37 was then attached to the *première Escadre de Chasse* (1st combat wing), part of General

Above: Portrait taken in the early 1920s.

Duval's air division. This great air armada was systematically sent to all the hot spots of the front and countered the German offensives. It was in this capacity that the SPA 37 fought in Picardy in March, on the Chemin des Dames in May, at the Battle of the Matz in June, at the Battle of Champagne in July, and the Battle of Ile-de-France in July and August.

Bernard Barny de Romanet, who had been granted two leaves of absence in September 1917 and from December 21, 1917 to January 5, 1918, began the year with a course at the Cazaux shooting school from January 8 to 31, 1918. He came out of this training course much better trained in aerial shooting... On March 20, 1918, he was on leave when the German army launched a major offensive on the Somme. He was urgently recalled and joined his squadron the next day to begin five months of intense fighting.

On March 31, 1918, he shot down an Albatros D fighter near Mondidier, in the Somme, in collaboration with two other pilots. He was promoted to lieutenant on May 9, 1918 (effective April 15) and on May 30, 1918, he won his third victory over a Drachen. He became an ace by shooting down two enemy aircraft on June 1st and 30th 1918, and increased his score with two more on July 16th and 17th 1918, which earned him the Knight's Cross of the Legion of Honour on July 30th 1918.

On August 22, 1918, he shot down his 10th enemy aircraft in collaboration with his squadron leader, Lt Roger Poupon.

Below: After his participation to the air meeting of Brussels, the top aces of French aviation are received by the king and queen of Belgium in their castle of Laeken. In the foreground, from left to right, Amand Pinsard, Bernard Barny de Romanet, queen Elisabeth of Belgium, Charles Nungesser, and René Fonck. Those 4 men totaled 163 victories…

Above: The SPAD XXbis tested in Etampes by Bernard Barny de Romanet on September 25, 1920.

He became famous because his name was mentioned by name in the communiqué to the army, which was published in the newspapers on August 25, 1918.

Commander of SPA 167

But by the time he could read his name in the newspapers, he had left the SPA 37 squadron. Young lieutenant of active duty and well noted by his superiors, he was considered as an elite officer and entrusted with the command of a newly created squadron, SPA 167, destined to become the 5th squadron of the n°12 fighter group, known as the Cigognes. De Romanet chose for his insignia a stork with raised wings, called "de Romanet", and set about making his unit an elite formation on a par with the other units of the GC 12, the SPA 3, 103, 73, and 26.

His natural authority seemed to give good results since his unit obtained a dozen certified victories during the last months of the war, himself obtaining 8 certified victories to bring his score to 18. His deputy officer, S/Lt Guiguet, an ace with 5 veteran victories from Guynemer's SPA 3, gave a qualified assessment of his leader: "*He came from a noble family with a racehorse stable in the East, and had a pilot brother. He was an active soldier who spoke very harshly to civilians, which I was never able to handle. But from a military standpoint, I recognize that he was very bright.*"

After the armistice came into effect on November 11, he was granted a long leave from the 18th to the 30th of that month. Back at his squadron, which was now stationed in occupied Germany, he was hospitalized on December 18, 1918, probably due to the Spanish flu epidemic, and was discharged on January 6, 1919, for a 10-day convalescence, followed by another leave from February 1 to 23, 1919.

Test pilot

He left the SPA 167 on March 1, 1919 to be assigned to the military mission in America, and upon his return a few weeks later, found a situation in the private sector to the point of requesting a two-year leave from the army on September 5, 1919.

He joined the Breguet company as a commercial advisor and left for Spain to deliver a Breguet 14 to the Spanish government with his fellow aviators Walbaum and Massot. He returned in April 1920 to deliver a new aircraft to the Spanish government, giving him the opportunity to fly from Paris to Madrid in eight hours.

But it was mainly as a racing pilot for the SPAD company that he became known to the general public, participating in several speed races in which he was in constant competition with the pilot Sadi Lecointe, who flew on Nieuport. The press enjoyed this competition of aviators and regularly reported on their recent exploits...

On October 22, 1919, de Romanet takes the world speed record from Sadi Lecointe by flying at 268 km/h during tests for the Deutsch de la Meurthe cup. On May 2, 1920, on a seaplane version of his aircraft, he faced Lecointe again but lost the race to an Italian Macchi led by the pilot Zanetti. Decorated with the Officer Cross of the Legion of Honor on June 16, 1920, he and Lecointe are the two aviators to finish the Gordon Benett Cup that the other competitors abandon... Lecointe is the winner but de Romanet took his revenge at the

Bernard Barny de Romanet Victory List								
N°	Date	Unit	Plane piloted	Plane claimed	Place	Notes	German casualties	Confirmation
1	3-May-17	N 37		Enemy Plane	Craonne			Citation OA 23-5-1917 (MM) - JORF 7-7-1917 p.5213
P1	12-Jul-17	N 37		Enemy Plane				
P2	17-Oct-17	N 37		Enemy Plane	SE Fort de Vaux	With Brig André Delannoy		
2	31-Mar-18	SPA 37		Albatros D	Rollot	With Lt Roger Poupon et MdL Marc Coupillaud		Citation OCA 21-4-1918.
3	30-May-18	SPA 37		Drachen	Fère-en-Tardenois	With MdL Jacques de Lostalot		Citation OA 13-4-1918 - JORF 1-8-1918 p.6671 "3rd victory"
4	1-Jun-18	SPA 37		Two-seater	Villers-Cotterêts			Citation OA 14-7-1918 - JORF 20 -8-1918 p.7340 "4th victory"
5	30-Jun-18	SPA 37		Fokker D.VII	Chouy	With Lt Roger Poupon	Lt Feige, Jasta 10, KIA near Chouy.	Citation OA - JORF 7-1-1919 p.229 "5th victory"
6	15-Jul-18	SPA 37		DFW C	Laval-sur-Tourbe			Citation OA (LH) "6th and 7th victory"
7	16-Jul-18	SPA 37		DFW C	Champagne	With Adj Georges Lienhart and MdL Georges Francisquet		Citation OA (LH) "6th and 7th victory"
8	9-Aug-18	SPA 37		Enemy Plane	Mortemer	With Lt Roger Poupon, S/Lt Fernand Guyou and MdL Marc Coupillaud		
9	15-Aug-18	SPA 37		Fokker D VII		With Adj Louis Marot and MdL Léon Daniel		
10	22-Aug-18	SPA 37		Two-seater	Soissons	With Lt Roger Poupon		
11	3-Oct-18	SPA 167	SPAD XIII	Fokker D VII	St. Etienne-à-Arnes	With Cprl Lombardon		Citation OA 19-9-1918
12	10-Oct-18	SPA 167	SPAD XIII	Two-seater	Bignicourt-Laneuville			General Order n°1470 28-11-1918 (12th, 13th, and 14th victories)

Bernard Barny de Romanet Victory List								
N°	Date	Unit	Plane piloted	Plane claimed	Place	Notes	German casualties	Confirmation
13	14-Oct-18	SPA 167	SPAD XIII	Fokker D VII	Les Allez-Bois de Voncq	With Lt Dumas		General Order n°1470 28-11-1918 (12th, 13th, and 14th victories)
14	18-Oct-18	SPA 167	SPAD XIII	Fokker D.VII	Civry-sur-Aisne	With Lt Dumas		General Order n°1470 28-11-1918 (12th, 13th, and 14th victories)
P3	18-Oct-18	SPA 167	SPAD XIII	Fokker D.VII	Civry-sur-Aisne			
15	23-Oct-18	SPA 167	SPAD XIII	Two-seater	S . Le Chesne	With S/Lt Le Chevalier		Citation OA - JORF 11-2-1919 p.1592 (15th, 16th, 17th, and 18th victory)
16	23-Oct-18	SPA 167	SPAD XIII	Fokker D.VII	Attigny			Citation OA - JORF 11-2-1919 p.1592 (15th, 16th, 17th, and 18th victory)
17	24-Oct-18	SPA 167	SPAD XIII	Two-seater	N. Attigny	With S/Lt Guiguet		Citation OA - JORF 11-2-1919 p.1592 (15th, 16th, 17th, and 18th victory)
18	29-Oct-18	SPA 167	SPAD XIII	Two-seater	S. Bois de Loges			Citation OA - JORF 11-2-1919 p.1592 (15th, 16th, 17th, and 18th victory)

Buc meeting by winning the world speed record at 292 km/h. Record that Lecointe takes back to him the following day at 296 km/h...

In February 1921, he goes to Spain to be the test pilot of the Pescara helicopter that the French government intends to buy. The tests are spread out over several months and de Romatet, which prolongs on June 5, 1921 his leave with the army for one year, returns in France to prepare the next race Deutsch de la Meurthe. Unfortunately, he crashed his plane on September 23, 1921 and was killed on the Etampes airfield.

All the personalities of French aviation, from the minister Laurent-Eynac to the great aces Fonck, Nungesser, and Pinsard, attended his funeral, with the King of Spain Alphonse XIII sending a telegram of condolences to his family. He was buried in Macon at the Saint Brice cemetery and a monument in his honor was inaugurated at Le Bourget by the minister Bokanowski in 1927.

Above: Air meeting of Buc. Bernard Barny de Romanet poses in front of his SPAD XXbis.

Facing Page, Above: President of the French Republic Alexandre Millerand shakes the hand of Bernard Barny de Romanet during the air meeting of Buc in October 1920, under the eyes of Ace Amand Pinsard (27 victories). The tall man behind the president is Representative Pierre-Etienne Flandin, then undersecretary of state for aviation.

Facing Page, Below: The SPAD XX bis of de Romanet was put on floats for the Monaco air race in 1921.

Left: Bernard Barny de Romanet (right) taken in photo with two attendants of the Buc air meeting in October 1920: Top ace René Fonck (left) and capitaine Gustave Lagache (center), commanding officer of escadrille SPA 3 in 1919.

Meeting de Monaco. Spad pilote
de Romanet 1921
34

Above: Bernard Barny de Romanet in front of one of his SPAD XX bis, date unknown.
Facing Page: Bernard Barny de Romanet found death at the command of this Lumière-De Monge racer on September 9, 1921.

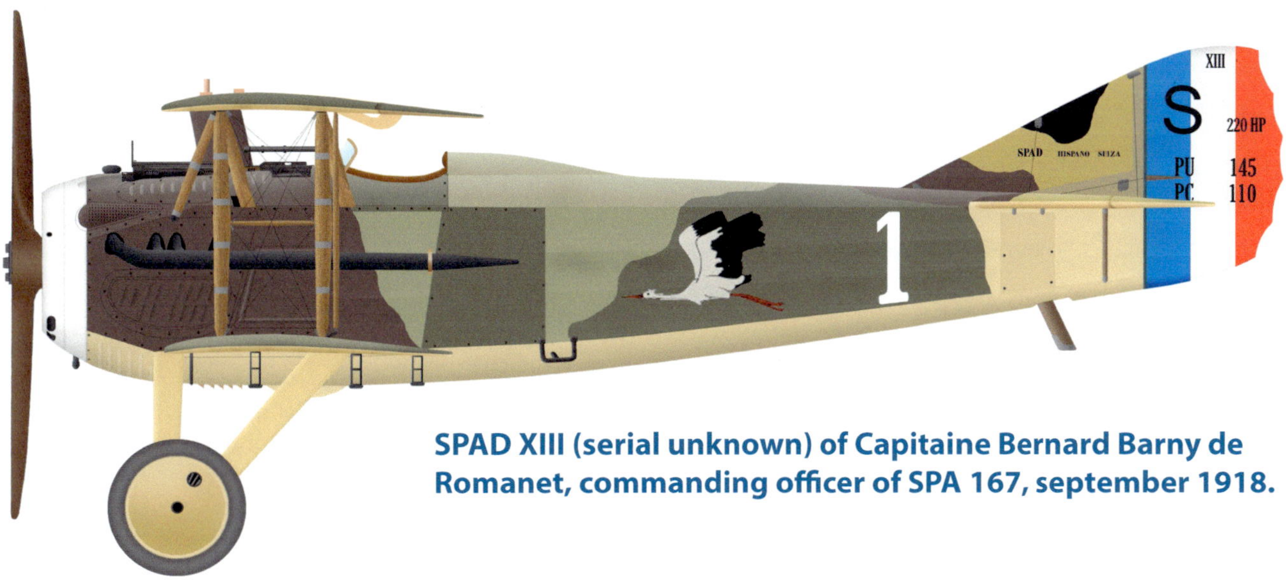

SPAD XIII (serial unknown) of Capitaine Bernard Barny de Romanet, commanding officer of SPA 167, september 1918.

François Battesti

First names: François, Marie, Noël.
Born: 05-05-1890 in Azzana (Corsica).
Father's profession: Notary.
Military pilot brevet: 03-04-1914 (N°453).
Units: BL 18, BL 3, C 10, SPA 73.
Army communiqué: No.
Total score: 7 victories, 1 probable.
Died: 24-08-1977 in Azzana (Corsica).

Above: Sergent François Battesti near a Bleriot XI in year 1914.

A pre-war active military pilot

François Marie Noël Battesti was born on May 5, 1890 in Azzana, a village in southern Corsica located in the interior of the island, 20 kilometers northwest of Ajaccio, and which had about 500 inhabitants at the time. His family was rather well-to-do since his father was a notary and the mayor was probably related to him since he had the same surname.

The young boy was visibly unfit for regular education and at the age of eighteen left for Ajaccio on October 14, 1908 to voluntarily enlist in the army for a period of five years. He was drafted into the 108th infantry regiment in Bergerac, Dordogne, as a private 2nd class. He served with dedication and slowly gained rank: he became corporal on September 26, 1910, then sergeant on October 19, 1912.

When his enlistment ended on October 14, 1913, he signed up again for the 24th infantry regiment, in the Paris region. It was in this unit that, fascinated by the emerging aviation, he volunteered for the aeronautical service where he was detached on December 3, 1913 and trained as a pilot, obtaining his military pilot's license (n°453) on April 3, 1914.

He was thus part of the very small phalanx of professional military pilots when the war broke out, and found himself assigned to the BL 18 squadron in Dijon on August 4, 1914, flying on Bleriot XIs known as "Channel Crossing". Transferred to BL 3 in Belfort, he flew with his mechanic Bonnard on an aircraft that was overloaded and crashed at takeoff, from a height of about 20 meters. Both men were injured and Battesti was seriously injured, disfigured by facial wounds and with a double fracture of the right leg (tibia and fibula).

Reconnaissance pilot and artillery spotter

He was discharged for several weeks of hospitalization and re-education, which ended on December 14, 1914, when he went to the Dijon depot. On December 27, he was sent to the Avord school to train on new types of aircraft, and on January 6, 1915, he went to the Pau school, where he met a young apprentice pilot named Georges Guynemer for the first time. On January 20, 1915, he left for the general aeronautical reserve in the Paris region and on March 20, 1915, he received his first assignment to an operational unit, the BL 10 squadron, which at that time abandoned its old Bleriot aircraft and converted to Caudron G3s.

Above: Affected to escadrille C 10, adjudant Battesti poses with his dog Jocko near his Caudron G4 in the airfield of Pierrefonds, between July 1915 and May 1916. The name "L'aiglon" (The eaglet) is a reference to old imperial France, as the eaglet was the nickname of the heir of emperor Napoléon the first, a man from Corsica as was Battesti.

Battesti spent almost two years of service there and was promoted to *adjudant* on May 9, 1915, flying first on a single-engine Caudron G3 and then on a twin-engine Caudron G4. He left no doubt about his origins and affinities by decorating his G4s with the nickname "*L'aiglon*" (eaglet) or "*Petit Napoléon*" (small Napoléon) ! These planes were often pierced by enemy flak, which he bravely faced on each adjustment mission, as underlined by his first citation in the army order on December 4, 1915, which awarded him the military medal.

These missions were not without danger and he was the victim of an accident on February 9, 1916 when he crashed with his plane, then more seriously on July 8, 1916 while flying over enemy lines at low altitude where he was hit by a shot that destroyed one of his engines. Losing altitude, he managed to return in extremis to his lines on an engine but crashed violently, being seriously injured by a fracture of the right collarbone.

He was evacuated to the military hospital in Abbeville and during his convalescence he learned of his promotion to second lieutenant on July 27, 1916. Returning to his unit on August 16, he volunteered to serve in a fighter squadron and his request was accepted in early 1917.

Above: Portrait of François Battesti, probably taken in early 1916 just after he received the médaille militaire.

Above: Battesi next to a Nieuport 17 from escadrille N 73, early 1917.

Fighter pilot at N 73

On February 2, 1917, he left for the training division group to perfect his skills in handling the Nieuport fighter. On March 12, 1917, he was assigned to an operational unit, escadrille N 73, a fighter unit whose command had just been entrusted to the ace Albert Deullin and which was part of the Combat Group n°12, known as the Groupe des Cigognes, an elite unit headed by Major Antonin Brocard.

Battesti was received by Brocard upon his arrival and told his commander that it was an honor for him to be assigned here, without his having asked for it. Brocard told him that here one either shoots down the Boche or is shot down, otherwise one is sent back to his former affectation!

He made his debut as a fighter on the Chemin des Dames, where he won his first certified victory on April 24, 1917, and then, most likely aboard a SPAD, won a second one on July 4, 1917, which earned him a promotion to the rank of lieutenant. The Stork Group then went to Flanders to support the British air force in the so-called Passchendaele offensive. The German air force faced the sector with its best units and the losses were heavy for the French fighters, which lost its greatest aces: Captain Matton was killed, Captain Heurtaux was wounded and put out of action for good, and Captain Deullin was injured. The worst loss for the morale of the pilots was the disappearance of the ace of aces, Georges Guynemer, on September 11, 1917. Battesti, who had flown several combat missions with him, would venerate his memory for the rest of his life.

Our Corsican pilot survived the fighting in Flanders and won a third certified victory on November 12, 1917, over the Houthulst forest. He was in fact the best pilot of the SPA 73, just after his commander Deullin who had taken over his post and who, at the beginning of 1918, upon hearing the news of the award of the Legion of Honor to his teammate, gave it to him personally by removing his own medal from his chest to give it to a very moved François Battesti.

Flying now on a SPAD XIII in 1918, he won new victories on May 22, June 12, and July 16, 1918, making him an ace but without any celebrity because the threshold to appear in the official communiqué was raised to 10 aerial victories. Nevertheless, a privilege reserved for aces, he received a SPAD XII-Canon in September 1918 with which he probably scored his 7th and last success on October 29, 1918, north of Laon.

Active officer with a temperament during the interwar period

Once the armistice was signed, Lieutenant Battesti wished to remain in the army to make a career of it. He was then seconded to the staff of the No. 1 combat wing in 1919, and on January 1, 1920, he was given command of the SPA 95 squadron stationed in occupied Germany.

The young victorious pilot during the war was then transformed into a frustrated and brutal soldier, as shown by the written notes left by his superiors. Col. de Vaulgrenant, head of the 1st air division, notes it in a very mitigated way: "*Magnificent war service. Very military character. Devoted, conscientious. But shows on all occasions a violence, an anger which makes his job rather difficult. Very estimable nevertheless.* "However, he also wrote: *"Very energetic officer, but violent temperament of Corsican. Superb war services. Has much to learn."*

General Duport, head of his corps, noted on November 15, 1920 "*an officer of average value, difficult to lead in peacetime.* "While another of his senior officers, Major Houdemont, indicated "*Violent, devoted, full of heart. Lieutenant Battesti had several incidents with civilians in the city, especially speaking German. Brutal and unmitigated commander. Rudimentary administrator.* "

Battesti, undoubtedly driven by the desire to fight, asked to be assigned to the Levant (Syria) Air Force where fighting took place against the natives. He was transferred on May 18, 1921 to the 35th aviation regiment of the Levant army aeronautics, but, affected by malaria, did not fly much although he was interested in reconnaissance and bombing missions at the 54th squadron which he joined on June 10, 1921. His transfer on August 20 to 54 Squadron of the Levant Army, where he commanded the "fighter" element, did not change anything. He was then sent back to France for medical reasons on October 30, 1921, and was assigned to the 3rd aviation regiment.

After a year in a metropolitan unit, he left for the Levant

Above: SPAD VII n°1243, probably a plane flown by Lieutenant Battesti.

at his own request and joined the 35th aviation regiment on October 1, 1922, where he was promoted to the rank of captain on June 25, 1924, and received the rosette of Officer of the Legion of Honor on August 13.

Back in France, he got married on March 11, 1926 with Miss Gabrielle Guilminot, with whom he had been living together for one year. In 1927, he volunteered to be a flight instructor in Romania for the local military. His request was not accepted, because his superiors judged that his lack of poise and education would give too unfavorable an image of French aviation...

Although he was still considered an outstanding pilot (he received a letter of congratulations from the Minister Laurent-Eynac for having led a group flight of fighters from Strasbourg to Warsaw from August 28 to September 5, 1929), his character problems did not improve; quite the contrary. His note for the year 1929, which only follows that of the previous years, indicates: "*Remarkable performer and valuable instructor, has a well-run and well-directed squadron. General training a little weak and character too impulsive. Could nevertheless command a fighter group under good conditions because of his personal drive, prestige, and experience.* "

He returned to the Levant in 1932 to be appointed deputy to the commander of the Aleppo group, who noted him "*Primary education and training. Very violent and susceptible character.* "The following year, General Bigot du Granut, commander of the troops in the Levant, wrote of him: "*I have sometimes noticed that the magnificent soldier that is Captain Battesti had his head close to his cap. But I pay full tribute to his superb record of service, to his uprightness, to his value as a leader and trainer of men. He will do wonderfully at the head of a hunting group.* "His disciplinary file thus mentions 10 days of rigorous arrest on March 7, 1934 for having answered his superior officer, during a visit of the base by a foreign general officer...

On May 17, 1934 he returned to France in a metropolitan unit. He joined Marcel Buccard's Francist party, which openly proposed to establish fascism in France on the model of Italy. The judgment of his new commander was in line with the previous ones: "In peacetime, complicates somewhat the task of command by the careful handling required by a character whose sudden changes can be explained by his serious war wounds."

On September 15, 1934, he was appointed as an instructor at the Istres school, training non-commissioned officers of the flying personnel. On April 5, 1935, he was the victim of an airplane accident that caused him a violent concussion. Recovering from the shock, he asked on June 15, 1935 for a permanent leave of absence from the flying personnel. On February 3, 1936, less than a month before his leave was to become effective, he had it cancelled, to the great displeasure of his superiors. Due to a shortage of officers, and despite his difficult character, he was kept in the flying personnel, promoted to the rank of commander on March 15, 1937 and posted to the Marrakech air base in Morocco, being promoted to Commander of the Legion of Honor on June 16, 1938.

Battesti and a lady. Considering the palms on Battesti's croix de guerre, the photo was taken in late 1918, probably just after the armistice.

Adrift during the Second World War

When the Second World War broke out on September 2, 1939, he asked to go to a combat unit. His chiefs were rather delighted to get rid of him and support his request... He was then assigned to the Levant Air Force, reaching Beirut on January 23rd 1940. He was entrusted with the command of the Levant n°3 group on June 5, 1940, then that of the Beirut base on July 4. On September 30, 1940, having reached the new age limits instituted by General Bergeret, the Vichy government's Secretary of State for the Air Force, Major Battesti was disbarred from the active army and placed in the reserves.

He was authorized to reside in the Levant, from where he approached his superiors to reintegrate into the active army. Particularly in June 1941, during the Anglo-Gaullist attack on Lebanon and Syria, he offered his services to join a fighting unit of the air force. His requests went unheeded...

Returning to France by boat like many French colonists, he settled in Marseille, then Aix-en-Provence, where he brooded over his resentment. He joined Jacques Doriot's Parti Populaire Français, a collaborationist party, and wrote several articles in "*l'émancipation nationale*", one of the party's publications. Putting his money where his mouth was, he participated in the hunt for STO (Compulsory labour service for Germany) draft dodgers

Right & Below: Just after the war, François Battesti, now capitaine, poses near a fighter Gourdou-Leseurre GL 32 wearing the symbol of escadrille Lafayette, around year 1930.

in his region from the summer of 1942.

On November 8, 1942, upon hearing the news of the Allied landing in North Africa, he immediately wrote to General Jannekyn, Secretary of State for the Air Force in the Vichy government, to ask to be mobilized in the Air Force to fight the invasion. He received a direct written reply from the latter on November 16, 1942, giving him a polite refusal.

Battesti answered him on December 5, 1942 with unprecedented virulence, reproaching him for not having mobilized him in 1941 in Syria, while another officer, "*one of your Freemason friends*", had indeed been activated. He also blamed him for the loss of North Africa:

"*It would have been necessary at the head of our glorious air force a leader, a trainer with the soul of a Guynemer, and not that of a sad politician, a great profiteer of the misfortunes of our country. It doesn't matter. There are still Frenchmen, pure ones, worthy of the Maréchal. They are numerous, more numerous than we think, and decide to end it. Also, I can affirm to you, Mr. Secretary of State for the French Air Force, that France will be saved in spite of you and all those who are like you.*

Who lives? France!

Signed: Battesti"

On December 16, 1942, Jannekyn replied that the government had never been in control of its decisions with the German Armistice Commission forbidding any delivery of reinforcements. He concludes: "*I will not respond, however, to your insults, they betray a lack of balance that it is painful for me to see in the old soldier that you are.*"

While the free zone was now invaded by the German army, Battesti was going to fall into the ultra collaboration by being accused in his file of "*having joined the militia or an equivalent organization.*" On August 5, 1944, he was accused of having participated in the looting of various farms in the Charleval region, northwest of Aix-en-Provence.

He disappeared from the region when free French and American troops landed on the south coast of France. A gendarmerie investigation dated March 1945 tells us that "*on August 15, 1944, the day of the Allied landings in the Mediterranean, Battesti, a notorious collaborator, accompanied by his wife and maid, left his home and left in a German car. Afterwards, he would have been seriously wounded or shot by patriots in the Lyon area.* "

Arrested at an undetermined date towards the end of the war, he was accused of having trained in Germany in a school of intelligence and sabotage. Charged with undermining the external security of the State, he was incarcerated in the Beaumettes prison in Marseille, from where he was released on February 6, 1946 at the request of the prison doctor to enter the Timone hospice reserved for the mentally ill.

On June 17, 1947, the Bouches du Rhône Court of Justice sentenced him to 20 years of hard labor, 20 years of disqualification, national degradation, and confiscation of property. The next day, he was removed from the air force reserves as a disciplinary measure.

He did not serve his full sentence, as he was probably amnestied like many collaborators after a few years in prison - on December 18, 1948, he was still interned at the social hygiene center in Marseille.

Also struck off the rolls of the Legion of Honor on April 5, 1954, he moved to Algiers and became a local representative of the Association des vieilles tiges (pioneers pilots). In 1958 he panicked the administration of the Air Force for having participated in a reunion of old-timers where a senior Air Force officer spoke. A message from Paris to the local authorities reminded them that ex-commander Battesti was forbidden from any activity in an air force association, and that he was strictly forbidden to enter the air bases.

When Algeria became independent, he probably returned to his native village of Azzana in Corsica, where he wrote a book of memoirs published in 1975, "*Les Cigognes de Brocard au combat*" (Brocard's Storks in Combat), in which he saluted the memory of his former comrades in combat from the First World War in a rather bombastic style. It is worth noting a chapter in which he pays tribute to Pierre Costantini, a fighter pilot in the French Orient Army who became an ultra-collaborative militant during the Occupation and was sentenced to 15 years of hard labor upon liberation. He died shortly after the publication of his book, on August 24, 1977, in the town of Azzana.

François Battesti Victory List							
N°	Date	Unit	Plane piloted	Plane claimed	Place	Notes	Confirmation
1	24-Apr-17	SPA 73	SPAD	Enemy plane	Ste Croix (Aisne)		Citation OA June 14, 1917 - JORF Sept. 29, 1917 p.7695
P1	3-Jun-17	SPA 73	SPAD	Enemy plane	Pontavert		
2	4-Jul-17	SPA 73	SPAD	Enemy plane	Berry-au-Bac		Légion d'honneur Jan. 5, 1918 - JORF Jan. 7, 1918 p.300 "shot down enemy planes on July 4th and Nov 12th"
3	12-Nov-17	SPA 73	SPAD	Enemy plane	W. Houthulst forest		Citation OA Dec. 10, 1917 - JORF Jan. 7, 1918 p.300 "3rd victory" -
4	22-May-18	SPA 73	SPAD XIII	Enemy plane	Biaches		Citation OA June 13, 1918 - JORF August 1, 1918 "4th victory"
5	12-Jun-18	SPA 73	SPAD XIII	Enemy plane			Citation OA July 17, 1918 - "5th enemy plane"
6	16-Jul-18	SPA 73	SPAD XIII	Enemy plane		With S/Lt Claude Fontaine	Citation OA Nov. 25, 1918 "two enemy planes"
7	29-Oct-18	SPA 73	SPAD XIII	Enemy plane	N. Laon		Citation OA Nov. 25, 1918 "two enemy planes"
German casualties: No data.							

SPAD VII n°1234 of escadrille SPA 73 flown by Lieutenant Battesti in 1917.

Frank Baylies

First names: Frank Leaman.
Born: 23-09-1895 in New Bedford (Massachusetts, USA).
Father's profession: Grain merchant.
Military pilot brevet: 20-09-1917 (n°8718).
Units: SPA 73, SPA 3.
Army communiqué: July 8, 1918, for 12 victories (posthumously).
Total score: 12 victories.
Died: 17-06-1918 in Arvillers (Somme) – Killed in aerial combat.

Above: Frank Leaman Baylies portrait.

Frank Leaman Baylies was born on September 23, 1895 in New Bedford, Massachusetts, to a rather well-to-do family whose father was a grain merchant who later became a merchant. The young boy grew up in the city of Providence (Rhode Island) and in late 1915 was a 20-year-old student, who while in New York heard a speech in favor of his country entering the war on the side of the allies.

He decided to enlist in the American Ambulance Service on February 26, 1916, and a few weeks later found himself evacuating French wounded on the Verdun front. Caught by German artillery, he was awarded a citation in the army corps order which earned him the Croix de Guerre. He was then transferred to the Eastern Front where a local pilot offered him his first flight at the Salonika air park. Enthused, Baylies decided to become a pilot even though his country had officially entered the war in April 1917.

After returning to France, he first tried to join the American Air Force, but was rejected for physical reasons. He then joined the French Air Force on May 26, 1917, via the Foreign Legion, and immediately went to flight school where he obtained his pilot's license in Avord on September 20, 1917, and the rank of corporal a few days later. Completing his training at the Pau acrobatics school, he was then directed on November 16, 1917 to his first operational unit, escadrille SPA 73 led by ace Albert Deullin.

He flew his first war missions aboard a SPAD XIII with the tactical number 13, which inspired him to make the comment "*I can't afford to be superstitious*" in a letter to his family. However, he did not stay long in this unit, as he was transferred on December 1, 1917 to the prestigious escadrille SPA 3, the famous Stork squadron where Georges Guynemer served. He arrived there with an American comrade from SPA 73 named Edward David Judd, who was soon taken by the American army then in training in France and which was very keen on experienced executives to whom it offered officer ranks. Baylies received such an offer, but declined it, preferring to remain at SPA 3 where he only had the rank of corporal.

He opened his score at the controls of his SPAD XIII on February 19, 1918, by shooting down a two-seater that was certified for him. He was thrilled with the victory, as he wrote to his family: "*It was very exciting, much better than duck shooting and much more profitable.* "The SPA 3 was emptied of its historic aces Guynemer, Dorme, Deullin, and Heurtaux, but a new

Above: Frank Baylies taken by photographers of the Army Photographic Service in Hétomesnil on May 8, 1918. A SPAD XIII is on the background.

generation of young pilots took over in 1918; many of whom would become aces, such as his compatriot Ted Parsons and the Frenchmen Benjamin Bozon-Verduraz, Louis Risacher, and André Dubonnet. Baylies proved himself to be an excellent pilot by winning two new victories in March, which earned him a promotion to the rank of sergeant.

The German offensive that began on March 21, 1918 increased the number of opportunities for aerial combat; SPA 3 found itself in the front line and even had to evacuate its Mesnil-St-Georges airfield (near Montdidier) on March 24, in the face of advancing enemy troops. Ted Parsons testifies that his comrade Baylies was on a dawn patrol that day, wearing his pajamas under his flight suit; when he returned to the field with his tank almost empty, he was surprised to discover a number of aircraft bearing German crosses. German troops had just taken over the field... He immediately turned on the throttle to take off, while two German soldiers clung to his wings, shouting at him to surrender. Baylies switched to the auxiliary tank and took off under the fire of the sentries, landing ten minutes later in the Allied lines. On March 28, 1918, he had another scare when he was shot down by the defensive fire of a two-seater he was attacking, and had to land in the no-man's land where he regained the French lines thanks to the fire of French soldiers who were firing on his pursuers.

Multiplying the fights, he won 2 new victories in April and shot down his 6th official success on May 2, 1918, a Rumpler C, which he commented on by saying that he "*allowed Mr. Kraut to benefit from two machine guns that worked perfectly and were well regulated. He had little to say and ... fell out of control, hit the ground with a terrific bang and deposited a crumpled mass of debris there.* "

The month of May would be a great time for him as he would win 6 more official victories, bringing his total to 12, and often teaming up with ace André Dubonnet. But on June 17, 1918, while leading a patrol with Dubonnet and Sergeant François Macari, he came across a patrol of four aircraft flying at higher altitudes that he identified as British fighters. They were in fact Fokker Triplanes of Jasta 19 led by the ace Rudolf Rieunau. The Germans attacked them and Baylies made the mistake of engaging these planes, which were inferior in speed but superior in maneuverability. Placing himself in the tail of a German, he was shot down by the ace Rieunau and crashed in flames in the town of Arvillers. Macari managed to escape in a dive, while Dubonnet brought back his aircraft riddled with bullets.

The German air force then sent a message on July 6, 1918 stating that "*The pilot Baylies was killed in action and buried with military honors.* "Two days later, the army communiqué mentioned his name and stated that he had won 12 victories.

Baylies in flying suit near his SPAD in Villeneuve sur Verberie airfield, April 1918.

Died for France at the age of 22, Frank Baylies' body was transferred in 1927 to the Lafayette Escadrille Memorial in Marne-la-Coquette.

Above: Frank Leaman Baylies portrait.

Above: Frank Baylies wearing his decorations. We can note he wears the médaille militaire, a decoration officially given to him… posthumously in 1922! Obviously, he wears decorations he was proposed for by his military hierarchy, but not officially confirmed to him by the authorities.

SPAD XIII n°2430 of Frank Baylies, escadrille SPA 3, Hétomesnil, May 1918.

Frank Baylies Victory List								
N°	Date	Unit	Plane piloted	Plane claimed	Place	Notes	German casualties	Confirmation
1	19-Feb-18	SPA 3	SPAD XIII n°5759	Two-seater	N. Forges			
2	7-Mar-18	SPA 3	SPAD XIII n°2433	Fighter	NE. Courtecon		Jasta 76, Ltn Moritz Weig POW.	Citation OA - JORF 10-6-1918 p.4996 "2nd enemy plane"
3	16-Mar-18	SPA 3	SPAD XIII n°2433	Enemy plane	Chevrigny		Jasta 79b, Ltn Fritz Edler von Braun WIA	Citation OA - JORF 21-7-1918 "3nd enemy plane"
4	11-Apr-18	SPA 3	SPAD XIII n°2433	Two-seater	Mesnil-St. George			
5	12-Apr-18	SPA 3	SPAD XIII n°2433	Fighter	S. Moreuil		Jasta 46, Ltn Aloïs Weber POW.	
6	2-May-18	SPA 3	SPAD XIII n°2430	Rumpler C	Assinvillers			Citation OA - JORF 18-6-1918 p.5262 "5th and 6th victory"
7	3-May-18	SPA 3	SPAD XIII n°2430	Two-seater	Montdidier	With MdL Dubonnet		Citation OA - JORF 18-6-1918 p.5262 "5th and 6th victory"
8	9-May-18	SPA 3	SPAD XIII n°2430	Halberstadt C	Braches-Gratibus			
9	10-May-18	SPA 3	SPAD XIII n°2430	Two-seater	Montdidier	With MdL Clément		
10	28-May-18	SPA 3	SPAD XIII n°8236	Fighter	Courtemarches			
11	29-May-18	SPA 3	SPAD XIII n°8236	Enemy plane	Etelfay			
12	31-May-18	SPA 3	SPAD XIII n°8236	Enemy plane	Montdidier	With MdL Dubonnet		Army communiqué 8-7-1918 - JORF 9-7-1918 p.5931 "12 airplanes"
Note: JORF on September 1, 1922 p.4173 awards him posthumously the médaille militaire, and talks about "8 German planes shot down during the campaign."								

Facing Page: A photo from the personal photo album of ace André Dubonnet – he was his wingman when Baylies was shot down and barely escaped the fight, hisp lane riddled with bullets. He never forgot his American brother in arms.

Amand Berthelot

First names: Amand, Jean.
Born: 11-07-1894 in Saint-Marc (Finistère) – Today, city of Brest.
Father's profession: Land owners.
Military pilot brevet: 18-03-1917 (n°5703).
Units: CRP, SPA 15.
Army communiqué: 14-10-1918, for 10 victories.
Total score: 11 victories (2 Drachen), 4 probables.
Died: 19-09-1961 in Paris (16).

Above: Portrait taken during the war.

A young Breton volunteer

Amand Jean Berthelot was born on July 11, 1894 in St. Marc, a commune in Finistère that is now merged with the city of Brest. His parents, who lived in the manor of Ker-Nic-Ti in Lambézellec, were undoubtedly well-to-do people who were able to offer their son an education. He became a young man of 1m74 with blue eyes and black hair, who obtained his baccalaureate in science and a fluent command of the English language.

On October 8, 1912, some time after celebrating his 18th birthday, Armand Berthelot decided to enlist in the army at the Brest town hall. Two days later, he was drafted into the 3rd Dragoon Regiment (RD) in Nantes as a private 2nd class. Six months later, on April 2, 1913, he was promoted to the rank of brigadier. He was still in his unit when the First World War broke out and left for the front on August 2, 1914, towards the Belgian border on the Meuse, before retreating and blocking the German troops by participating in the Battle of the Marne. The 3rd RD was then sent to the Yser during the "race to the sea" where it held the front for several months, before being sent at the end of the year to Lorraine, where, divided into "half-regiments", it held the front as a simple infantry unit.

Brigadier Amand Berthelot, promoted to the rank of maréchal des logis on April 2, 1915, received a citation in the order of his regiment on September 7, 1915, for "*on November 12 and 13, 1914, in front of Boësinghe, [having] ensured day and night liaison with neighboring infantry troops violently attacked and crossing a zone heavily battered by artillery.*"

Pilot at the CRP in 1917

He was probably not satisfied with the life of an infantryman in the trenches that the evolution of the war made him lead and asked for his transfer to the air force. This was granted to him on October 3, 1916 when he left for the Dijon-Lonjvic school to follow his theoretical training as a pilot. He then moved to the Etampes flying school on November 5, 1916 where he obtained his military pilot's license (n°5703) on March 18, 1917. He then moved to the Avord training school on March 23, and then to the Pau acrobatic school on April 28, 1917.

He soon received his first assignment to a squadron and according to his military file he joined N 365 on May 22, 1917, a unit in the Camp Retranché de Paris (CRP, entranched camp of Paris), but which does not exist in the squadron nomenclature - this may be a copy error in his file. If Amand Berthelot did join the CRP, he hardly had the opportunity to meet the enemy, who rarely showed up in the capital. Parisian aviators were frequently treated as "hiders" by pilots coming

Above: Photo printed in a newspaper, showing a SPAD XIII of ace Amand Berthelot (in the cockpit) with little tombstones on the crest of the fuselage, used as victory markings. The plane also shows knight Bayard's helmet on the side, the symbol of escadrille SPA 15. Tactical number (if any) and serial number are unknown.

Right: Portrait taken after the war, showing his legion of honor he received after the armistice.

from the front... But because he did not face the enemy, Berthelot gained valuable additional flying experience during his first weeks in the squadron, which would be crucial when he went to the front.

According to his military file, he officially left for the front on July 27, 1917 – the day the Germans carried out their only raid on Paris in 1917 – and jointed escadrille SPA 15, which was grouped with other units in the GC 13. However, another source mentions his presence for four months at the CRP, which is very plausible because according to the logbook of SPA 15 his first flight dates only from October 18, 1917.

Whatever the exact date of his arrival, the atmosphere was quite different in this new squadron that was stationed in the Meuse, on the Verdun front, during the fall of 1917. The unit, whose symbol was the Bayard helmet, had many aces, including adjudant Gabriel Guérin (6 victories as of October 18, 1917, out of a total of 23 at the end of the war) and especially S/Lt Lucien Jailler (12 victories), who had the reputation of taking young pilots under his wing to teach them the tricks of the trade.

Above: Photo showing another plane of the ace, with the tactical number "2" on the crest of the fuselage. Note the two red and white stripes on top wing.

Among top aces of SPA 15

His first results were not long in coming. On October 24, 1917, while flying with Sgt. Artigau and Sgt. Le Corneur, his patrol discovered the presence of 5 Albatros D III single-seaters heading towards a French observation balloon. Artigau was the first to shoot at one of them, followed by Berthelot, who cleared an AR two-seater attacked by one of the fighters. His target was seen to leave the fight in a disoriented dive, and was not certified.

The two victories he claimed on the following December 29 were also not certified: during a patrol in the Reims region, he attacked an Albatros D III at noon in a patrol of three and fired 20 rounds at it. The enemy fighter dived vertically towards its lines, pursued by Berthelot who tried to disengage its machine gun. The Albatross straightened up at 1200 meters and Berthelot, who had disengaged, fired again. The German fell in a wood but, caught by enemy flak, could not determine where he had fallen. Later in the same patrol, he shot a DFW at 12:30 above Tahure but could not finish him off because his machine gun jammed – his prey was seen stabbing into his lines in an unnatural way, the rear gunner appearing to be killed.

His first certified successes came during the German spring offensives, where air activity was increased tenfold. On March 24, 1918, he shot down a Drachen in flames at 5:00 p.m. in the NE of the St Gobain forest, which earned him his promotion to adjudant on April 1 and his first citation in the Army Order on April 16, 1918. But by then he had already achieved a second success, burning down another Drachen at Mézières on April 12 at 4:40 p.m., under the protection of the other pilots in his patrol.

Other successes were to follow quickly, as he and two other pilots forced an enemy two-seater to land in the French lines on 15 May 1918. On May 27, 31 and June 9, 1918, he shot down three DFW C two-seaters, bringing his record to 6 victories, which earned him the military medal. A photograph shows that he marked his victories on the back of his SPAD XIII, with a tombstone design for a destroyed aircraft...

His hunting record continued to grow: on August 14, 1918, he shot down a Rumpler C near Chavigny with barely 10 rounds of ammunition, then a new enemy aircraft on September 4, 1918, while taking part in numerous ground attack missions on the retreating German troops. He shot down another DFW C on 26 September 1918, and finished his record with a "double" by shooting down two enemy fighters near Bantheville, his 10th and 11th victories.

He gained a certain amount of glory from this last feat, as he was mentioned in the communiqué to the army dated October 14, 1918, which was reproduced in the national press where several titles published a short biography of the pilot. However, he ended the war without being awarded the Knight's Cross of the Legion of Honor, which he received only on April 25, 1919.

Resigns from the army to re-enlist for Syria

Remaining mobilized at the SPA 15 after the armistice, he remained an active soldier and was assigned on April 2, 1919 to the Service Technique de l'Aéronautique, to the flight test department at Villacoublay. There he was promoted to the rank of second lieutenant and made numerous flights on various types of aircraft, including a flight from Paris to Bordeaux in jut two hours on a SEA IV built by Henri Potez.

He left the army on October 23, 1919 by resigning his rank and retired to Lambezellec, his parents' home in Finistère.

Amand Berthelot Victory List								
N°	Date	Unit	Plane piloted	Plane claimed	Place	Notes	German casualties	Confirmation
P1	24-Oct-17	N 15		DFW C				
P2	29-Dec-17	N 15		Albatros D	Butte de Souain - Ferme de Medevils			
P3	29-Dec-17	N 15		DFW C	Tahure / Monthois			
1	24-Mar-18	N 15	SPAD	Drachen	NE. Forêt de St Gobain			Citation OA 16-4-1918 - JORF 13-5-1918 p.4161 "1 Drachen in flames"
2	12-Apr-18	N 15	SPAD	Drachen	Mézières			Citation OA 22-4-1918 - JORF June 25, 1918 p.5471 "2nd balloon"
3	15-May-18	SPA 15	SPAD	Two-seater fighter	Gamelin	With Lt Laporte, MdL Kuhling. Crew captured		Citation OA 30-6-1918
4	27-May-18	SPA 15	SPAD	DFW C	Comin	With Adj Artigau		Citation OA 18-6-1918 - JORF July 24, 1918 p.6381 "4th victory"
5	31-May-18	SPA 15	SPAD	DFW C	Grottes			Citation OA 22-9-1918 - JORF Jan.3, 1919 p.284 "5th victory"
6	9-Jun-18	SPA 15	SPAD	DFW C	Bois de Chaux - Oulchy le château			Citation OA (MM) 7-7-1918 - "6th victory"
7	14-Aug-18	SPA 15	SPAD	Rumpler C	Chavigny			Citation OA 27-9-1918 - JORF 3-1-1919 p.278 "7th and 8th victory"
8	4-Sep-18	SPA 15	SPAD	Enemy plane		With Sgt Fourcade, Sgt Barat		Citation OA Sept 27, 1918 - JORF Jan.3, 1919 p.278 "7th and 8th victory"
P4	26-Sep-18	SPA 15	SPAD	Two seater	Villers-devant Dun			
9	26-Sep-18	SPA 16	SPAD	Fokker D .VII	Montfaucon			Confirmation buy decision of GC 13 of 1-10-1918
10	9-Oct-18	SPA 15	SPAD	Enemy plane	Bantheville			Citation OA 7-12-1918 - "10th victory"
11	9-Oct-18	SPA 15	SPAD	Enemy plane	Bantheville			Citation OA 28-12-1918 - JORF 5-7-1919 p.6889 "11th victory"

His activity during the following years is not well known: he is mentioned as a participant in the aviation meeting in Caen on August 21 and 22, 1920, and indicates in a correspondence that he worked for the Franco-Romanian company while carrying out flight tests in Villacoublay. However, his situation in civilian life did not seem to suit him, as he was looking to return to the army. As he could not return to service with his rank of reserve lieutenant, he resigned it and re-enlisted on August 30, 1922 as a simple adjutant pilot with the 35th RAO in Bron, with the aim of being posted to Syria where the French air force was fighting. He was given satisfaction and landed in Beirut on October 18, 1922, being assigned to the 59th squadron of Muslimié, then one week later to the 55th squadron of Rakka and from November 10, 1923 to the 1st squadron of Rayak. He would serve a little less than two years in the Levant with application and a great enthusiasm, being very well noted by his superiors.

An active soldier without motivation

His colonial period ended on April 24, 1924 with his return to Marseille, from where he joined the 34th RAO (Aviation Observation Regiment) at Le Bourget after spending a four-month leave with his family in Finistère. He returned to his squadron after this long period of rest, but the enthusiastic young pilot of the war seemed to have changed a lot. Described as having an "embittered and worried" character, his direct superior, Major Houdry, noted about him at the end of 1924 that: "*For the past six months, adjudant Berthelot has completely betrayed the trust placed in him. His private life is deplorable. He does nothing to reduce his debts, he lives, it seems, only on expedients, and to give the change has recourse to unworthy lies. It is absolutely no longer possible to make a proposal for him to be given the rank of officer.*"

Another of his superiors, Lieutenant Julien Guertiau, also a former ace of 14–18 (8 victories), did not share this opinion and found that adjudant Berthelot had given complete satisfaction in all respects. But he changed his mind at the beginning of 1925, describing his subordinate as very irregular in his way of serving, disinterested in his service and riddled with debts.

However, his new unit commander at the end of 1925 noted that he was an excellent pilot who regretted his past misdeeds. He supported him in his promotion to officer, allowing him to get his second lieutenant's stripe again in March 1926, then some time later his temporary lieutenant's stripe while serving in the 6th fighter squadron of the 38th Aviation Regiment. His father having died at the end of 1925, he requested a transfer near Brittany to solve his inheritance problems.

His improvement in the way he served was obviously a flash in the pan, since his commanding officer, Colonel Estirac, noted him in 1928 as "*a mediocre officer, making little effort for his general or military training, flying little and irregularly, living on a past as a brilliant fighter pilot.* "Often on leave due to health problems, he was pushed out of the army and retired on August 15, 1930 with the rank of reserve lieutenant.

Special assignment in 1940

A married man with no children, he reconverted to civilian life by becoming an insurance inspector. However, he did not leave aviation: he attended the dinner of the association of ace pilots chaired by Fonck in 1934, the same year he resumed training in the reserve by performing several periods of voluntary exercises. He was considered an excellent pilot, being placed in class A, suitable for flying fighter planes.

Promoted to Officer of the Legion of Honor on July 30, 1937 and Reserve Captain on June 25, 1938, he was mobilized again in September 1939 when the Second World War broke out. He was quickly given a special assignment at Hispano-Suiza, where, with the help of two lieutenants and a company of colonial infantry stationed at Maisons-Laffitte, he was in charge of the military and air defense of the underground Hispano factories in Houilles and Bois-Colombes.

Fully committed to the Resistance

He withdrew to the south along with the company during the debacle, and was demobilized in Pau on June 18, 1940, a few days before the armistice came into effect. But if France was to come out of the war, Captain Berthelot refused to capitulate and was one of the few Frenchmen to immediately resist against the enemy.

On June 21, 1940, he organized the departure for North Africa of 19 aircraft from Pau airfield. On November 2, 1940, he returned from the exodus with the Hispano-Suiza workers, who were resettling at their Bois-Colombes plant. Realizing that the plant would only work for the enemy war effort, Berthelot resigned on November 9, 1940, but organized a group among the workers to delay the delivery of engines and spare parts to the enemy, which remained active throughout the war.

Thanks to the support of friends, former pilots of the First World War, he found a management job at the Pelletier et Fourrures company at 36 bis avenue de l'Opéra in Paris on July 2, 1941. But even in this position he was confronted with the fact that he was working for the enemy war effort: he was involved in making clandestine seizures of stocks allocated to the German army and distributing clothes to prisoners of war.

As the months passed, the resistance became more structured and Amand Berthelot organized a team to collect and transfer parachuted Allied airmen to Spain. In June 1942, he set up a medical team to provide immediate care to Allied aircrews who had landed on the Normandy coast, in the Livarot region. The Gestapo dismantled the organization in September 1943 by arresting and deporting the doctor.

Affiliated with the "Liberation" movement, he collaborated with the resistance movement "les ardents" of the SNCF (French national railway company), as well as with the Anglo-French liaison. Thanks to the complicity of the Bois-Colombes town hall administration, he had nearly 200 food cards distributed to STO draft dodgers, escapees, and other people wanted by the Germans.

In 1944, Berthelot became the commander of the FFI of Bois-Colombes and took an active part in the liberation battles in the area which started on August 16, 1944. The men under his command were able to seize the town and secure the camp on the Gennevilliers plain, where there were important depots of military equipment. Berthelot personally took two prisoners during the fighting, in addition to the 17 others captured by his men. While the Germans took 18 hostages (including 3 women) to cover their retreat on the north bank of the Seine at the Bezons bridge, Berthelot obtained their release by courageously negotiating during the night of August 22–23, 1944, threatening to execute 50 German prisoners for each hostage kept in reprisal.

On January 4, 1945, FFI commander Amand Berthelot was recalled to active duty at the Nanterre air force warehouse. He served there with the rank of lieutenant-colonel for almost a year, but despite his request to be kept on active duty and the support of his superior, he was sent home by decision of the Minister of the Army on February 13, 1946.

A dispute over stripes

Officially demobilized on May 7, 1946, he retired to Colombes and became a sales representative. General de Gaulle awarded him a citation to the order of the air force on December 15, 1945, which was followed by another one on May 13, 1947, to the order of the army corps, for his activities in the resistance. He was also appointed reserve colonel on May 25, 1946, decorated with the Resistance medal on August 3, and made commander of the Legion of Honor on August 24.

But his post-war career was marked by a dispute with the military administration... He was accused of illegally wearing a military rank by having proclaimed himself a commander, a rank he had never obtained in the reserves under Vichy. The military administration launched a procedure, which was quickly abandoned by the hierarchy.

On April 16, 1947, the commission for the liquidation of the Resistance's regulations was faced with the problem of validating his rank, since he was already a reserve colonel... Faced with a fait accompli, the commission validated his promotion to the rank of reserve lieutenant-colonel on September 25, 1944, and then colonel on May 25, 1946, but demanded that he repay the overpayment of his salary, which he refused to do, starting a procedure that he lost.

Colonel Amand Berthelot was disbarred from the Air Force on November 9, 1956 because of the age limit and died on September 19, 1961 at the age of 65.

SPAD XIII (number unknown) of Amand Berthelot, 1918

Maurice Bizot

First names: Maurice, Félix, Léon.
Born: 05-11-1896 in Puéchabon (Hérault).
Father's profession: Mechanic.
Military pilot brevet: 09-07-1917 (n°7370).
Units: N 90.
Army communiqué: No.
Total score: 10 victories (7 Drachen), 1 probable.
Died: 27-11-1925 in Versailles (died in hospital after a flying accident).

Above: Portrait.

Mechanic from Grenoble who volunteered for the army

Maurice Félix Léon Bizot was born on November 5, 1896 in Puéchabon, a town in the Hérault region of France, where his father worked as a mechanic. The family then moved to Villard-Bonnot in the Isère region, near Grenoble, where the young man, who reached the height of 1m68, followed in his father's footsteps and became an electrician's apprentice.

When the First World War broke out, he had not yet celebrated his 18th birthday and therefore had not done his military service, which at that time was at the age of 20. Nevertheless, caught up in the general patriotic fervor, he decided to enlist voluntarily with the authorization of his father on January 8, 1915 by presenting himself at the town hall of Grenoble. He was immediately assigned to the 11th foot artillery regiment, stationed in the city.

After completing his training, he was sent to the front and his good education level earned him a promotion to the rank of brigadier on April 6, 1915. Transferred to the 3rd foot artillery regiment on March 5, 1916, he moved to the 82nd heavy artillery regiment on July 1. Then came nearly 29 months of service in the artillery at the front, without any particular brilliance, but also without receiving the slightest wound.

Probably discouraged by a service in the rear of the front, Brigadier Maurice Bizot volunteered to serve in the air force. His request was officially accepted on May 18, 1917, when he was detached to the 1st aviation group and began his pilot training. After starting his theoretical training at the Dijon school, he moved on to the Istres flying school where he obtained his military pilot's license (n°7370) on July 9, 1917, on a Caudron G4. He then went on to the Avord advanced training school, and then to the Pau acrobatics school where he was trained to fly fighter planes. After a spell at the Groupement des Divisions d'Entrainement du Plessis-Belleville, he was sent to his first operational squadron, the N90 stationed at Toul, which he joined on October 13, 1917.

Fighter pilot

The squadron, whose symbol was a rooster, was in a rather quiet sector of the front and did not have the priority of the general staff as far as its equipment was concerned, since it

Facing Page: Photo of Maurice Bizot (3rd from left) likely taken during early summer 1918, with other pilots of SPA 90. 2nd from left is ace Charles Macé (12 victory), killed in a flying accident in 1919, who shared with Bizot his first two victories in March 1918. On the right is ace Jean Pézon (10 victories) who also shared 3 victories with Bizot. Contrary to what is written on the photo, Pézon died in his bed at very old age.

Above: Photo taken in escadrille SPA 90 in Germany just after the armistice in 1919. Bizot, on the center under the propeller of the SPAD XIII, wears all his decorations except his légion d'honneur given to him in 1920.

Facing Page: Maurice Bizot in 1920, employed as test pilot in the Blériot Company.

was still flying Nieuport 24, 24bis, and 27 fighters, which were under-engined in the face of the German two-seaters, which they were struggling to catch up with. Nevertheless, among its members, there was a team of young and very reckless pilots, consisting of S/Lt Marc Ambrogi, Sergeant Charles Macé, and Corporal Jean Pezon who, despite this handicap, soon distinguished themselves by multiplying their successes, especially against the German captive balloons, the Drachen.

Maurice Bizot began to distinguish himself and earned a first mention in dispatch for having forced to return in his lines a German observation plane one November 5, 1917, and also to have carried out bombing missions by dropping grenades on enemy airfields and depots.

On March 12, 1918, with Corporals Pezon and Sullivan as comrades, he claimed a two-seater that was shot down near Château-Salins, but which was not certified. His official hunting record began on March 27, 1918 with a "double": two enemy aircraft shot down between Lagarde and Parroy, in collaboration with adjudant Charles Macé and adjudant Maurice Ruamps, a SPA 87 pilot.

As N 90 received its first SPAD fighters in April 1918, it is most certainly on board one of these aircraft that he would set fire to his first Drachen in company of S/Lt Ambrogi on July 30, 1918 near Grange-en-Haye, but his contribution to this victory seem to have not been confirmed. The enemy balloon was in itself an easy target... but very dangerous to approach, because it was always protected by strong flak that easily sifted with lead any reckless fighter daring to approach it.

This did not make Maurice Bizot give up, and he set fire to two others in quick succession on August 7 and 8, 1918, his 4th and 5th personal successes, even though he had been promoted to the rank of adjudant on the 1st of the month. He set fire to another one with S/Lt Ambrogi on August 21, and then shot down with two comrades a two-seater LVG of FA 281, whose crash near Autrepierre is confirmed in German archives.

On September 2, 1918, he won a "double" by burning 2 Drachen of the Balloonzug 62 and 36, the first with SLt. Dencausse, and the second in collaboration with Lt. Ambrogi and Adjutant Pezon. But luck deserted him because this time he did not get through the flak barrage, as he confided to the journalist Jacques Mortane: "*My greatest fear was the day when I was wounded while returning from setting fire to a Drachen. I was 200 meters*

83247

Above: Maurice Bizot (left) posing in Le Bourget airfield in 1924 with capitaine Jean Dagnaux and pilot Albert Villechanoux, behind a Blériot 115 bis prepared for the last two men for a raid in Africa.

high, under the violence of the shock, I don't know what I did, I found myself in a spin. Was I going to crash to the ground at that moment? I stand up half-blinded by blood, I see in front of me a line of trees. I have the impression that I am going to hit it. What happens inside me is indescribable. I close my eyes, believing myself irremediably lost. An agonizing minute! And suddenly, I found myself in normal flight, passing the lines very low, greeted by the Boche machine guns. I was saved. Fear, joy, and what emotion! "

When he returned to the field, he was awarded the military medal for his exploit, which forced him to stay in hospital until 18 September, after which he was granted a rest leave until 13 October 1918. He then returned to his squadron for the last battles of the war, during which he set fire to two new Drachen in collaboration on October 21 and 29.

Ending the war with a total of 11 certified victories, he would have been entitled to be quoted in the communiqué to the armies as is customary for any fighter pilot after his 10th success. Unfortunately for him, the news of his 10th success did not reach the general headquarters until after November 11, when the communiqué ceased to appear with the armistice. He did not get any fame for his fighter skills other than a short article in Jacques Mortane's specialized magazine, "*la guerre aérienne illustrée*".

Chief pilot at Blériot

Remaining in his squadron after the end of the fighting, he participated in the occupation of the Rhineland with it until September 22, 1919, when he was demobilized shortly after being promoted to the rank of adjudant-chef. He then declared that he was retiring to Bellegarde in the Ain region, but quickly moved to Versailles a few months later, as he found a job as a pilot instructor at the Blériot company in Buc.

He contributed to the training of several hundred pilots until 1923 when he became the company's chief pilot. In parallel to his civilian activity, he remained a reserve military pilot, assigned to the 3rd fighter regiment and then to the 35th RAO. He was made a knight of the Legion of Honor on July 11, 1920, and was promoted to the rank of reserve second lieutenant on July 31, 1923.

On November 27, 1925, Maurice Bizot flew over the Buc airfield on a competition SPAD. During a low pass, his propeller hit a mound and the plane went down at more than 200 km/h. Ejected from the aircraft, Bizot saw his fall cushioned by his parachute opening a few meters above the ground. But this was not enough: severely concussed, he was rushed to a hospital in Versailles, where he died a few hours later.

Above: Maurice Bizot flying a SPAD XV prototype.

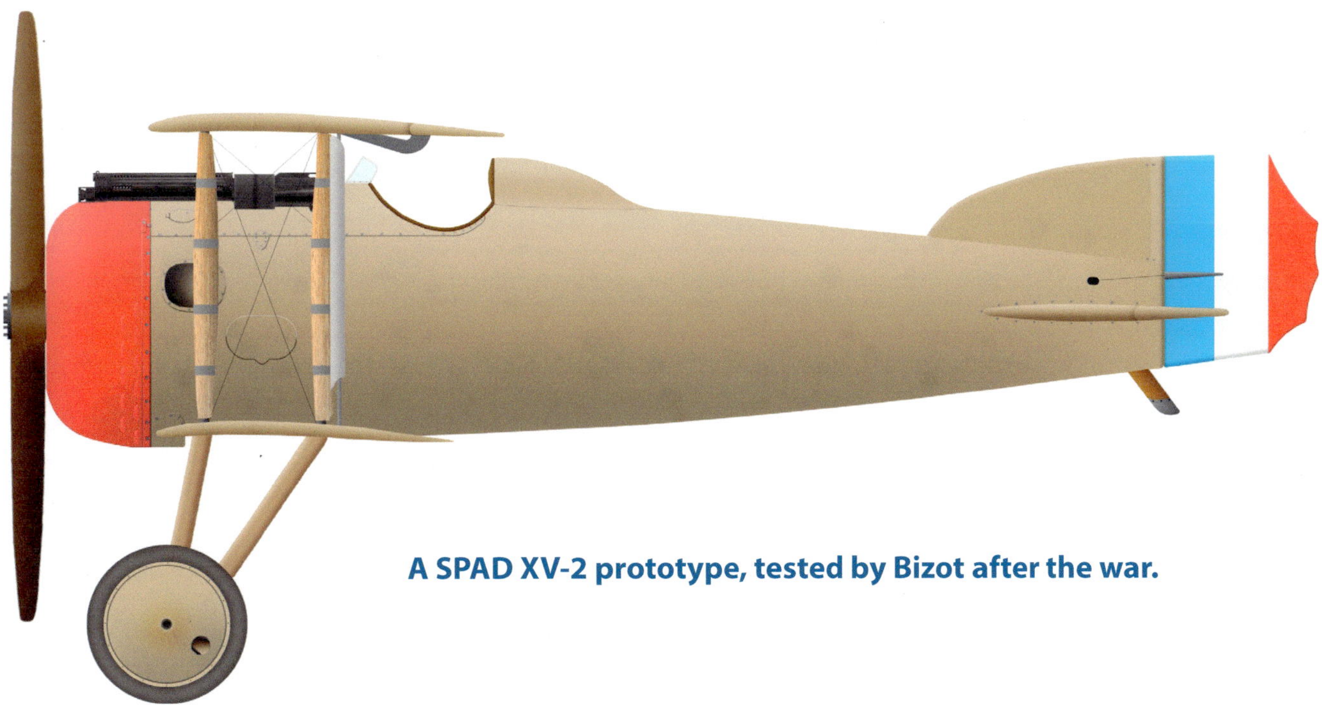

A SPAD XV-2 prototype, tested by Bizot after the war.

Maurice Bizot Victory List

N°	Date	Unit	Plane piloted	Plane claimed	Place	Notes	German casualties	Confirmation
P1	12-Mar-18	N 90	Nieuport	Two-seater	Château-Salins	With Cprl Sullivan and Cprl Pezon		
1	27-Mar-18	N 90	Nieuport	Fighter	Lagarde / Parroy	With Adj Macé and Adj Ruamps (SPA 87)		
2	27-Mar-18	N 90	Nieuport	Enemy plane	Lagarde / Parroy	With Adj Macé and Adj Ruamps (SPA 87)		
-	30-Jul-18	SPA 90	SPAD	Drachen	Grange-en-Haye	With S/Lt Ambrogi		Participation not confirmed
3	7-Aug-18	SPA 90	SPAD	Drachen	Juville			Citation OA 24/08/1918 - JORF 08/10/1918 p. 8755 "a Drachen in flames"
4	8-Aug-18	SPA 90	SPAD	Drachen	Vaxy / Amelecourt			Citation OA 31/08/1918- JORF 09/12/1918 p.10596 "an enemy Drachen"
5	21-Aug-18	SPA 90	SPAD	Drachen	Vaxy	With S/Lt Ambrogi		Citation OA 05/10/1918- JORF 03/01/1919 p.280 "an enemy Drachen (3rd victory)"
6	23-Aug-18	SPA 90	SPAD	LVG C	Autrepierre	With S/Lt Lemaire and Sgt Bizonnade	FA 281	Citation OA 10/11/1918 - JORF 20/01/1919 p.770 "23 august an enemy plane (…) 2 september a Drachen (…) 5th and 6th victories"
7	2-Sep-18	SPA 90	SPAD	Drachen	Geline	With S/Lt Dencausse	Balloonzug 62	Citation OA 10/11/1918 - JORF 20/01/1919 p.770 "23 august an enemy plane (…) 2 september a Drachen (…) 5th and 6th victories"
8	2-Sep-18	SPA 90	SPAD	Drachen	Juvelize	With Lt Ambrogi and Adj Pezon	Balloonzug 36	Citation OA (MM) 29/09/1918 "On 2 September, shot down 2 drachen in flames"
9	21-Oct-18	SPA 90	SPAD	Drachen	Avricourt	With S/Lt Bordes and Adj Pezon	Balloonzug 217	Citation OA 10/07/1919
10	29-Oct-18	SPA 90	SPAD	Drachen	Geline	With Adj Pezon and Cprl Alliot	Balloonzug 62, damaged.	Citation OA (LH) 12/07/1920 - JORF 12/07/1920 p.9856 "Shot 3 planes and 7 balloons"

Above: Maurice Bizot in front of a Blériot 115bis transport plane.

Above: Maurice Bizot (right) posing with famous raid pilot Georges Pelletier-Doisy, also WW1 ace, who became famous after making a Paris-Tokyo flight in 1924. They are taken in photo in January 1925 just before a raid Paris-Tchad prepared by Pelletier-Doisy in a Blériot 115 bis that we see on the background with his personal insignia, a cat standing in a crescent moon.

Below: "I'm a wimp, I don't fly any more. But bring me victories, and I'll give you medals !", said Captain Weiss (7th left), commanding officer of SPA 90, to his young pilots. They did! Many aces among them. Adjudant Maurice Bizot (1st left, 10 victories), Adjudant Pezon (4th left, 10 victories), Sous-Sieutenant Marc Ambrogi (8th left, 14 victories), and Adjudant chef Charles Macé (2nd right, 12 victories).

Pierre-Georges Blanc

First names: Maurice, Félix, Léon.
Born: 18-06-1887 in Châtignac (Charentes).
Father's profession: Farmer.
Military pilot brevet: 31-12-1915 (n°2232).
Units: C 47, N 31, SPA 154.
Army communiqué: No.
Total score: 5 victories (1 Drachen), 1 probable.
Died: 25-04-1960 in Béziers (Hérault).

Above: Portrait of Pierre-Georges Blanc (1887–1960).

A farmer's son

Pierre-Georges Blanc was born on June 18, 1887 in the Charente village of Châtignac, 20 km south of Angoulème. He came from a modest background where his father was a simple farmer; the witnesses signing his birth certificate were another farmer and the village teacher.

The young boy, who would grow up with his sister Andréa, thirteen years his junior, will lose his father at the beginning of the century. His mother took over a farm in the village of Sainte-Marie. Pierre-Georges, who became a sturdy man of a rather imposing size if one believes the photographs, sometimes appearing simply under the first name of Georges in the administrative documents, would leave the communal school very early without obtaining or even probably attempting the certificate of studies. He would become a salesman and on April 25, 1908 would get married at the town hall of Sully-sur-Loire with Miss Clémentine Reculé.

He then left to do his military service on October 7, 1908 and was drafted as a private 2nd class in the 131st infantry regiment stationed in Orléans. Two years later he was discharged with a certificate of good conduct and moved to Paris where he worked as a mechanic-electrician.

His marriage was apparently in trouble because he was noted as being single when the First World War broke out. He was then incorporated into the 50th infantry regiment of Périgueux on August 2, 1914 and fought with his unit in Belgium, then on the Marne before being sent to Champagne, then to Artois until July 3, 1915 when he was accepted on his request in the air force where his skills as an electrician-mechanic would undoubtedly be better used.

Beginning in reconnaissance aviation

After passing through the Dijon center for his theoretical training, he was then sent to the Avord school to take his flying lessons. There he received his pilot's license (n°2232) on December 31, 1915 on a Nieuport aircraft, and was promoted to the rank of corporal on January 10, 1916. From February 2, 1916, he was in the general aviation reserve (RGA) awaiting his assignment to an operational unit, a priori a fighter squadron.

But fate decided otherwise. On February 23, 1916, during a training flight, he missed his landing and crashed, breaking his plane, from which he was taken out seriously injured, suffering from a concussion and multiple contusions. He spent several months in care and only joined his first frontline unit on July 2, 1916.

This one was the C.47, an artillery-spotting unit equipped

with the twin-engine Caudron G4. A decision that may seem surprising for a pilot trained in the Nieuport fighter... But he was 30 years old at the time and perhaps no longer considered young for this profession. He joined his unit stationed in the Oise region, where the future ace of aces, Corporal René Fonck, was serving, having already won one certified victory and two probable ones. Perhaps the latter's example pushed him to seek out combat with the enemy? In any case, promoted to sergeant on August 26, 1916, he was attacked by three German fighters on October 20. His plane was riddled with bullets, but he fought bravely and, as some sources suggest, allowed his observer to shoot down one of the attackers, but the citation he received at the Army Corps level does not mention any victory. At last, he managed to return with his plane riddled with bullets...

Pierre-Georges Blanc left C 47 on February 13, 1917 to be hospitalized in Châtellerault. The reason for this stay is unknown, but probably was due to the after-effects of his fall the previous year, from which he kept a fracture in his right leg that was never fully recovered.

Passage to a fighter escadrille

He left the hospital fairly quickly, as he joined the N 31 squadron on April 2, 1917, stationed in the Marne. He flew his first fighter missions there and on the 15th of that month he left for the GDE to get a personal aircraft, probably a Nieuport.

With this new aircraft he quickly won victories. On May 6, with two others pilots, he attacked a group of two-seater aircraft, one of which was shot down east of Coury, near Reims, earning him a citation in the Army Order. Less than a month later, on June 2, 1917 at 9:50 a.m., he and a wingman forced a German two-seater to land in the French lines, where it was captured with its crew.

Appointed adjudant on June 29, he received a new palm on his war cross and left his unit for the GDE on July 14 to be trained to fly the SPAD VII on which he would be converted. But his recurrent health problems led him to be hospitalized again during the training period, and he only returned to his unit on July 18, 1917. He was granted a leave of absence from August 3 to 16, and resumed the fight, enjoying a period of success in October.

On the 7th of that month, in collaboration with S/Lt Dousinelle, he shot down an aircraft between Lierval and Laval. On the 17th, it was a Drachen that he set on fire with adjudant Chartoire over Luzilly. Then on October 30, he shot down a Jasta 34b fighter alone. With five confirmed victories he was now an ace and should normally be mentioned in the communiqué to the armies, as is the tradition. However, he had to pay the price for the new policy of the General Staff, which now reserved this honor for pilots who had obtained 10 certified victories - his former comrade René Fonck was the last pilot to be distinguished with 5 victories, on June 3, 1917. Maurice Boyau, the next pilot to be mentioned in the press release, won on October 11, 1917 for his 10th victory.

Pierre-Georges Blanc will never know the honors of the media coverage of the aces, although he would be decorated with the military medal for his series of victories on November 20, 1917. He did not have the opportunity to increase his score: benefiting from a two-week leave at the end of the year, he returned to the SPA 31 on January 2, 1918 and went back to the RGA from January 16 to 31. On February 12, he was transferred to SPA 154, another squadron of GC 11, in order to make it more dynamic. His new squadron leader, Captain Lahoulle, successfully introduced his pilots to Drachen hunting. Some of them would obtain important successes, first of all Lt Coiffard, who will reach the score of 34 victories. Pierre-Georges Blanc, for his part, only stayed three months in the squadron. Undoubtedly tired by the fighting, and undermined by the after-effects of his fall from the plane, he left the front

Above: Portrait of Pierre-Georges Blanc (1887–1960).

Facing Page: Pierre-Georges Blanc (center-right, with glasses and turning head) and Sgt Adrien Chapelle (center-left) a posing in front of the German two-seater they just shot down and captured on June 6, 1917, near Nanteuil-la-Fosse, NE of Soissons. On the left and the right of the two French pilots, stand the two German aviators just captured. The one on the right does not seem too happy to be a prisoner of war…

Pierre-Georges Blanc Victory List								
N°	Date	Unit	Plane piloted	Plane claimed	Place	Notes	German casualties	Confirmation
-	30-Oct-16	C 47	Caudron G4	Enemy plane	German lines	With an unknown observer		Citation OCA 26-12-1916. No mention of victory.
1	7-May-17	N 31	Nieuport	Two-seater	E. Courcy	With Cprl Chapelle and Sgt Roland	FA 286, Ltn Wilhelm Höding et Ltn Alfred Küffner killed.	Citation OA 09-05-1917 - JORF 29-09-1917 p.7692 "Enemy plane"
2	2-Jun-17	N 31	Nieuport	Two-seater	Nanteuil-le-Fosse	With Sgt Chapelle	FA 257, crew captured	Citation OA 05-07-1917 - JORF 04-11-1917 p.8810 "2nd enemy plane'
3	7-Oct-17	N 31	SPAD	Enemy plane	Lierval-Laval	With S/Lt Dousinelle (N 48)		
4	17-Oct-17	N 31	SPAD	Drachen	Luzilly	With Adj Chartoire		Citation OA 19-10-1917 "already shot down 3 planes (…) On October 17 flames a Drachen"
5	30-Oct-17	N 31	SPAD	Enemy plane	Moulin-Didier		Jasta 34b, Ltn Anton Warmuth killed in Filain	Citation OA 24-11-1917 "5th victory"

Above: Photo from the French magazine "*La guerre aérienne illustrée*» showing pilots of SPA 154 in early 1918. Many future aces: standing, 5th from Left, is capitaine Lahoulle (10 victories), and 7th is Jacques Ehrlich (19 victories). Pierre-Georges Blanc is standing on the right; tired with combat fatigue, he will soon leave the unit and the front to finish the war as a ferry pilot.

on May 12, 1918 to become a reception pilot at the SFA (Airplane Production Service) in Buc.

Invalid of war

Probably demobilized shortly after the armistice, he moved to Béziers where his health problems caught up with him. His fall on January 23, 1916, caused him a double fracture that never fully healed, resulting in a reduction of one and a half centimeters in his right leg. He was then proposed for reform n°1 (temporary) with a 30% pension by a medical commission on February 24, 1920, the same year he was awarded the Knight's Cross of the Legion of Honor but by minister's choice, without mention in dispatch, on November 9 (page 17927 of Journal Officiel).

In 1921, although he was assigned to the reserve in the 31st aviation regiment, he was transferred to the auxiliary service on the proposal of the reform commission, with his disability pension maintained. At the end of the year, he was transferred to the 3rd group of aviation workers, and in spite of his request to be maintained in the armed service, he was struck off the air force personnel for physical unfitness on August 1, 1923. Placed without assignment in the reserve in 1927, he was released from all military obligations in 1936 and did not serve in the armed forces during the Second World War. He was made an officer of the Legion of Honor on July 14, 1950, and died on April 25, 1960 in Béziers.

Marcel Bloch

First names: Marcel, Léopold.
Born: 21-07-1890 in La Chaux de Fonds (Switzerland).
Father's profession: Unknown.
Military pilot brevet: 31-12-1915 (n°2232).
Units: N 3, N 62.
Army communiqué: 02-10-1916, for 5 victories.
Total score: 5 victories (5 Drachen), 3 probables.
Died: 29-03-1938 in Brno (Czech Republic).

Above: Portrait of Marcel Bloch in a Nieuport 11 equipped with Le Prieur Rockets.

An international youth

Marcel Léopold Bloch, the namesake of the famous French aircraft manufacturer, was born on July 21, 1890 in Switzerland, in the border village of La Chaux de Fonds, north of the city of Neuchâtel. His parents, Lazare Bloch and Florine Lévy, lived in France, in the commune of Chaux in the Belfort region, and gave their son French nationality. As he grew, he reached the impressive height of 1m80 for the time.

The young man studied engineering in Switzerland and in 1909 joined the design office of the Antoinette Company, building the first airplanes. In 1910 he left for the United States to work for the Moisant Aeroplane Co and Queen Aeroplane Co (Loening), where he conducted engine and aircraft tests. In this capacity, he was sent to Mexico and Panama, and was seconded to the Washington military center to carry out demonstrations. In 1913, he was in Argentina to study the creation of aerial postal links, then moved to Brazil where he helped to form the Sao Paulo aviation school.

During this stay abroad, the French military administration asked him to do his military service with the class of 1911. He obtained a deferment of call-up until 1913 but at the end of this one he was declared insubordinate by the military administration on January 4, 1914.

Joined France at the declaration of war

When the war broke out in France, he immediately took the necessary steps to join France where his parents lived in St-Etienne, but the difficulty of finding a ship meant that he was only able to make himself available to the military authorities on September 7, 1914 after having crossed the Atlantic at his own expense. He was threatened with punishment but managed to explain his situation and was struck off the list of insubordination on September 15, 1914; he was then assigned to the 1st aviation group in Pau as a private 2nd class.

He was then assigned as a mechanic to the Pau aviation school which reopened in October 1914, where he adjusted aircraft. Promoted to corporal on March 14, 1915, he was accepted to be a student pilot and obtained his military pilot's license (n°2571) on October 12, 1915, supplemented by the civilian license (n°2561) which was granted to him the following November 12.

Debut in Guynemer's N 3 squadron

Directed to the general aviation reserve on November 21, 1915, he was assigned to the N3 squadron on January 19, 1916 and flew his first mission on the frontline at the controls of the single-seater Nieuport 10 n°266 on the 25th of the month over Ribecourt and Lassigny, in the Oise region. He then flew two-seater Nieuports (n°502, 465, 480) on board which he carried out army reconnaissance missions with an observer from March 1st. On March 13, he flew again on Nieuport 10 n°147 and met an LVG which he could not intercept, as his engine

Above: Portrait of Marcel Bloch in a Nieuport 11 equipped with Le Prieur Rockets.

was failing. Alternating fighter and reconnaissance missions on two-seaters, he was promoted to the rank of sergeant on April 11, 1916. All his missions earned him a first citation in the army order: "*Excellent pilot. Since his arrival in the squadron, he has made numerous flights over the enemy. Has shown great qualities of courage and coolness. Has fought six air battles. During a reconnaissance, on March 6, had as well as his observer the frozen face. Nevertheless, continued his mission until the end.* "

The first balloon-buster specialist

On May 25, 1916, the N 3 squadron stationed at Cachy split to form a new fighter squadron, the N 62, which Sergeant Marcel Bloch joined on his Nieuport 11 n°901. In this unit, he was going to take advantage of the shooting course he had done at Cazaux on Le Prieur rockets from February 1 to 22, 1916 when he was at N 3: he was going to specialize in hunting Drachens, the German captive balloons.

N 62 was stationed at Cachy, in the Somme, where a major British offensive was being prepared. On June 5, 1916, it took off with MdL Sanglier, both aboard Nieuport 16s armed with Le Prieur rockets. They were covered by six other aircraft of their squadron flying as protection. They each attacked a balloon: Sanglier missed his while Bloch, flying at 50 m altitude, set his Drachen on fire with his rocket salvo. Taken as a target by the flak, he dived towards the German lines and managed to return to his field. The air victory would not be certified to him...

On June 26, 1916, when the artillery preparation for the offensive began, the destruction of the Drachen became a very important objective. The four French squadrons grouped together at Cachy went on the hunt for this objective; Marcel

Above: Portrait of Marcel Bloch in a Nieuport 11 equiped with Le Prieur Rockets. The photo was taken in the airfield of the Nieuport-Astra factory, in Issy-les-Moulineaux, south-west of Paris.

Facing Page, Above: Being an aces gives many privileges... The company Hispano-Suiza puts at his disposal one of its cars, in exchange for this advertising photograph.

Facing Page, Below: Drawing of Marcel Bloch done by the lady he was dating, the famous artist Geneviève Granger (1877-1967).

Bloch did not find a target but nevertheless used his rockets, as he testified to the journalist Jacques Mortane: "*On June 26, 1916, I left in search of a Drachen, I went down to 50 meters above the place that had been indicated to me, but no matter how hard I looked, I did not see the balloon that had been announced. On the other hand, from the nearby airfield, three airplanes rushed up to pursue me. I did not hesitate, I fired my rockets against them and the fright that this jet of flames produced was enough to put them to flight.*"

A new attack patrol took place on June 29: "*Three days later, while going to hunt for Drachens, I experienced a strong disappointment; I found none. All had remained cautiously on the ground. On the way, I arrived at Peronne, where I was greeted by a very violent cannonade. I had only one resource left to stop it: I pretended to have been hit by simulating a fall. The batteries stopped firing and I reached 200 meters above the ground, I was 10 km inside the enemy lines. Suddenly, I see a bivouac, I observe the comings and goings of the troops. I can see a huge forage store. I do not hesitate. I go down again, and, at 50 meters, I launch my rockets. Immediately a fantastic fire breaks out. Without worrying about me, the soldiers rush to put out the fire. This is what I was waiting for, and at that moment, it is my machine gun that continues the work of the rockets. I shoot among this multitude and unroll my bands in all tranquility. And I returned without incident, despite the cannons, the rifles, the machine guns, flying at 100 meters from the ground. My aircraft was riddled, but I was safe and sound.* "

L'ADJUDANT BLOCH

Dessin de Geneviève GRANGER

Above: SPAD VII used by Marcel Bloch in late 1916, showing the first version of N62 insignia, a rooster.

The attack by British and French troops was launched on July 1, 1916, the day Sergeant Marcel Bloch was again on the attack with his Le Prieur rockets: "*July 1, 1916, was the day of the attack. I left, I attacked a first Drachen and succeeded in shooting it down despite the intervention of numerous enemy planes rushing after me, the enemy airfields being close by. I returned without being bothered. All was well. I take back other rockets and I leave in search of another captive balloon, I discover one. When I was about to set off my rockets, a German aircraft that had been chasing me came between the Drachen and me. He shoots frantically to prevent me from intervening and riddles me with bullets. I turned slightly and retaliated with a strip, while standing vertically at 400 meters from the ground. The plane is hit. It stings, seems to fall to crash, but I don't have time to follow it. The main thing is that it left me the road free. I continue my descent towards the Drachen and I have the chance to set it on fire in its turn. Yes, that day, I certainly felt the strongest emotion of my life. It was in Gueudecourt.*"

What Bloch did not specify in his testimony was that he was targeted by a cloud of gunfire that pierced his aircraft, which, hit in the engine, flew and lost altitude. At that moment, he was 30 meters high... He managed to land in the British lines where he noticed that, in addition to the fact that his fuselage was pierced on all sides, a piece of shrapnel had sliced through one of the wing struts! The two Drachens were confirmed.

On July 3, he was back on the attack with three wingmen on a new aircraft, the Nieuport 16 n°1254, while the French troops succeeded in breaking through the German lines: "*The sausage that I had to attack was at a relatively high altitude. But as soon as it saw me coming, it was shot down very quickly. I gave chase and reached him at 500 meters. I prepare to launch my rockets to pierce her side, when a bullet goes through my thumb. At that moment I had my hand on the contact. Under the influence of pain, I press in an instinctive gesture and, of course, I miss the goal. The Drachen continues its descent. I follow him closely, I suffer terribly, but I have to take revenge. I move to the side and at the moment when I prepare to attack him for the second time, a new bullet penetrates my thigh. Ah! I don't feel the pain anymore, the fury invades all my being, I made too much and especially one made me too much so that I give up. I insist, I must be victorious. I unroll a whole bunch of incendiary balls and I am lucky enough to see the ball burst into flames. I can now think of myself. The loss of blood causes several successive faintings. My aircraft, which was 50 meters above the ground, was riddled with bullets, and I managed to reach our lines by a miracle, I do not know yet how. I landed behind the British trenches, after having been a target for all the different weapons of the enemy.*"

He landed in the British trenches at Lahoussoye, and would have declared to the soldiers who had come to his aid *"It's done, I saw him burn"* before fainting. He was taken back to his squadron in a pitiful state: "*Towards evening, they came to get me in a car. I tried to resist, I started to eat, but the pain got the better of my energy and I was forced to be evacuated to a hospital where I stayed for two months. The Military Medal comes to reward me.*"

Above: This photo shows the second version of the SPAD VII *"Fantasio II"*, with the definitive version of N62's rooster – an angry one! Marcel Bloch didn't score any victory in this plane.

Ace of the communiqué

He was hospitalized in Amiens and remained in convalescence for nearly two months, during which time he was promoted to the rank of adjudant on August 16, 1916. He returned to his squadron at Cachy on September 1, 1916, and resumed his flights, probably on a new Nieuport 17. The day after his arrival, he attacked a Drachen which was not certified.

On the other hand, on September 30 and October 1, 1916, he would officially increase his record of achievements: "*On September 30, I succeeded in shooting down my fourth sausage in flames. I returned once again transformed into a skimmer.*

On October 1st, I go hunting again. According to the custom, I stand very low. This is not without some risks, because it is a thick brown and I am constantly afraid of bumping into the cables of our balloons. Near Bapaume, I see a sausage standing at 250 meters. I approach, of course. And at the same time, I count four others very close to each other. Each one is defended by guns, flaming caterpillars that go up vertically and planes. All of them converge on me. It's a big fight. You have to play finesse and bluff like in poker. I rush towards the first one, but before I get to it, I make a slight turn and dive at full speed on the second one, which, caught off guard, falls in flames before being able to try to retaliate.

I return to the field, after having made my way through the projectiles of all kinds. I land, do not even get off my plane and, while new rockets are being put in, I take a glass of port in a hurry, then I leave. I arrive at the same place. Nothing more. Carefully the other four Drachens were lowered to the ground. However, I don't want to have been disturbed for nothing. I always go down more and more, I sail at ground level in search of the fugitives. Suddenly I see one on the ground. I prepare to attack him. The Krauts realize that it is for him, so they do not hesitate to fire at me from above, at the risk of hitting him with their projectiles. They don't care if they damage him, because they think that I will get through if I am not shot first. I thought my last hour had come. I thought I was lost. Shrapnel was hitting my engine. Two cables were cut, a spar broken, wing struts severed. However, I managed to set my victim on fire and to return to the trenches, only 20 meters from the ground. I managed to land immediately afterwards, still in the English lines. I had triumphed over all possible attacks which had ended in a daring but futile pursuit of enemy aircraft. Following this flight, I had the joy of seeing my name mentioned in the communiqué, I was happy with these magnificent victories."

He is indeed mentioned in the communiqué to the armies of October 2, 1916 at 3 p.m., as having shot down 5 captive balloons to date.

He became famous, because the French newspapers greeted the arrival of the new ace and published a short biography in their pages, his photo being even published on the front page of the daily newspaper "*Le Journal*". The specialized journalist Jacques Mortane published an interview in his magazine "*La guerre aérienne illustrée.*" The magazine "*L'univers israëlite*" celebrates his co-religionist who has obtained the honors of the press release.

On October 26, 1916, he was in Paris to take delivery of a SPAD VII that he christened "*Fantasio*". However, insufficiently recovered from his previous wounds, he returned to hospital on November 2 and was struck off the rolls of his squadron on December 31, 1916.

He enjoyed several months of rest, during which the

Marcel Bloch Victory List								
N°	Date	Unit	Plane piloted	Plane claimed	Place	Notes	German casualties	Confirmation
P1	5-Jun-16	N 62	Nieuport 16	Drachen	Somme			
1	1-Jul-16	N 62	Nieuport 16	Drachen	Gueudecourt			Citation OA 27-07-1916 - JORF 25-03-1917 p.2376
2	1-Jul-16	N 62	Nieuport 16	Drachen	Gueudecourt			Citation OA 27-07-1916 - JORF 25-03-1917 p.2376
P2	1-Jul-16	N 62	Nieuport 16	Drachen	Gueudecourt			
3	3-Jul-16	N 62	Nieuport 16 n°1254	Drachen	N. Frise (Péronne)			Citation OA (MM) 14-08-1916
P3	3-Sep-16	N 62		Drachen				
4	30-Sep-16	N 62		Drachen	Longavesnes			Citation OA 17-10-1916
5	1-Oct-16	N 62		Drachen	E. Bapaume	October 2, according to citation		Citation OA 18-10-1916 - Communiqué 02-10-1916

newspaper "*Le Figaro*", always at the forefront of worldly affairs, announced in its March 8, 1917 edition his engagement to Miss Geneviève Granger (1877–1967), a renowned sculptor and ceramist. Jacques Mortane ironizes in his review: "*This young girl, whose works are exhibited in the Luxembourg, has a magnificent subject for a painting: Cupid, tired of his archaic arrows, asks adjudant Bloch to replace them with his incendiary rockets. The pilot accepts and after having shot down so many Drachens in fire succeeds in setting fire to a heart which communicates its flame to his. Ah! That I am not a poet!*:

Disabled in an accident in Russia

Promoted to second lieutenant on a temporary basis on February 11, 1917, Marcel Bloch was in fact designated to leave for Russia to join Commander Berger's military mission. He declared to the journalist Mortane that his ambition was to shoot down a Zeppelin. He left on March 23, 1917 and his journey, made by sea via Murmansk, was quite long. He joined the French pilots there, but his operational activity was quite short. On May 8, 1917, he took off from Anatra on a Nieuport fighter on which he made his first flight. He then fell victim to the first cause of loss for Russian pilots: the deficiency of their own equipment. The plane was seen to tip over on the wing and crash, Bloch was seriously injured in the face and had to be trepanned.

Very weakened, he was repatriated in September 1917 and remained for many months in convalescence after a stay in hospital. During this period he separated from Miss Granger and married Miss Geismar in Paris on June 25, 1918. Almost cured, he was awarded the Knight's Cross of the Legion of Honor on July 13, 1918, then sent on September 10, 1918 to the military mission in the United States from which he returned on March 1, 1919. He flew a total of 557 hours under the flag, of which 55 were in school, 405 at the front, 45 in Russia, and 52 in the USA.

Marcel Bloch had to reintegrate into civilian life by returning to his profession as an engineer and finding a job with the Gnome et Rhône company, but with a serious handicap caused by the after-effects of his 1917 accident, which caused him numerous dizzinesses and eye problems, not to mention joint problems in his legs. In 1922, the Seine reform commission proposed a temporary disability pension of 80%, which was reduced to 50% the following year and increased to 100% in 1934.

As a reserve officer, he was promoted to the rank of lieutenant in 1924, then captain in 1932, and decorated with the Commander's Cross of the Legion of Honor by Commander Dagnaux in 1935. The following year, he was promoted to Commander from a contingent of war-disabled soldiers at 100%.

He travelled a lot because of his profession and the official missions he carried out during the interwar period, going to Mexico from 1924 to 1926 to carry out aerial propaganda flights, as well as to Italy from 1928 to 1929 where he worked on the development of the Jupiter engines of the Dornier Wal seaplanes. At the same time, he was a member of aeronautical charities and in particular was secretary general of the mutual aid society for mutilated airmen, "Les ailes brisées" (the broken wings). He continued to fly and logged another 225 hours in Mexico and 32 in Italy.

While visiting the Zbrojovka establishments, he died on March 29, 1938 in Brno, Czechoslovakia, of meningitis. His body was cremated on the spot and repatriated to France to be buried in the cemetery of Marne-la-Coquette.

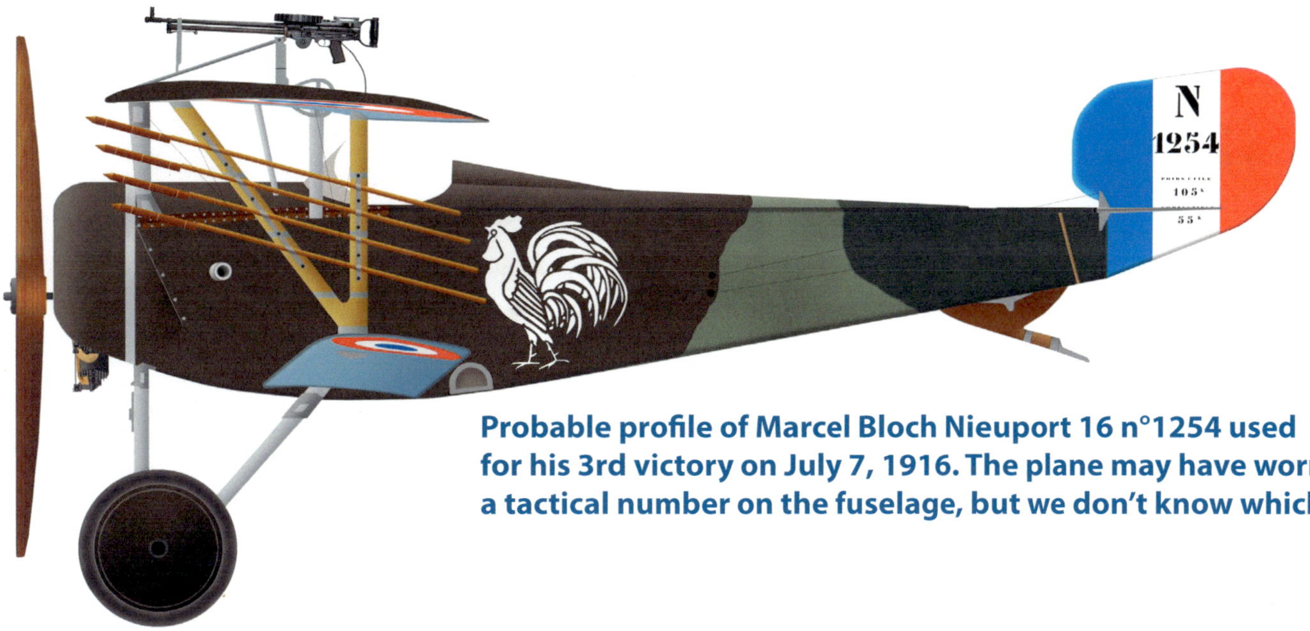

Probable profile of Marcel Bloch Nieuport 16 n°1254 used for his 3rd victory on July 7, 1916. The plane may have worn a tactical number on the fuselage, but we don't know which.

SPAD VII n°1?? of Marcel Bloch, Escadrille N 62, probably in Grivesnes, late December 1916.

François De Boigne

Name: de Bonald (Viscount).
First names: François, Eugène, Marie, Antoine.
Born: 18-08-1896 in Noyon (Oise).
Father's profession: Army officer (lieutenant).
Military pilot brevet: 25-08-1916 (n°4340).
Units: N 82, SPA 69.
Army communiqué: No.
Total score: 6 victories, 2 probables.
Died: 23-08-1970 in Neuilly-sur-Seine (Hauts-de-Seine).

Above: François de Boigne with his uniform of the 9th Hussars regiment, taken during his time in flying school in August 1916.

A son of a great aristocratic family

François, Eugène, Marie, Antoine de Boigne was born in Noyon, in the Oise region, on August 18, 1896. Contrary to appearances, his family ties were in Savoy, in the town of Chambéry: he came from an illustrious family of the department's nobility, having produced generations of notables. The founder of the dynasty was his great-great-grandfather, Benoist Le Borgne (1751–1830), who became Benoît Le Boigne, a Savoyard adventurer who made his fortune in India, where he became a general in the Mahratte Empire and married an Indian woman, with whom he had several children. He then settled in England during the French Revolution, repudiated his Indian wife and married a daughter of the French nobility, Miss Adèle d'Osmond, with whom he had no children and from whom he soon became separated. Returning to France during the Consulate, he settled in Chambéry with his important fortune in the castle of Buisson-Rond located in the city. Monarchist supporter and honored by Louis XVIII, he financed many charitable institutions in Savoie and particularly in Chambéry where one of the main streets bears his name. His second wife, the Countess of Boigne, also made a name for herself in literature by publishing her memoirs which painted an acidic portrait of France under the Restoration...

He was succeeded as Count of Boigne by his son Charles Alexandre (1791–1853), then by his son Charles Marie Ernest (1829–1895), who became a congressman and Mayor of Chambéry. This prominent man had two children, a daughter Edmée (1853–1937) and a son, Elzéar de Boigne born in 1865, the father of François de Boigne. Marrying a daughter of the aristocracy, Miss Hélène de Mandat de Grancey (daughter of a count), he decided rather than live off the family fortune, to make a career in the army. He was a cavalry lieutenant in the 9th cuirassiers in Noyon when François was born in 1896, then a captain in the 1st cuirassiers regiment in Alençon in 1901 when his second son Jean was born (1901–1995).

Unfortunately, a tragedy struck the family: Captain Elzéar de Boigne died following a bad fall from a horse in 1902, when François was only six years old. This death did not leave the young boy and his brother destitute, if one is to believe the impressive list of aristocratic names related to him that appear on the funeral notice. His mother, known as the Countess Elzéar de Boigne, who resided at the Château du Buisson-rond in Chambéry, remained very active in the social life of the time and the press reported all her travels on vacation, including the monarchist daily *L'Action Française* to which she was a faithful subscriber.

She was also able to offer a careful education to her children. When the war broke out in August 1914, François was an 18 year old student who had not yet fulfilled his military obligations. The service is normally done after the 20th birthday, so he was not mobilized.

Voluntary enlisted at the end of 1914

Nevertheless, he decided to enlist voluntarily to take part in the fighting, by going on November 11, 1914 to the recruitment office of the Chambéry town hall, with his mother's authorization. He was then incorporated as a simple soldier of 2nd class in the 9th regiment of Hussars where he followed his classes, the military administration noting that he was 1m72 tall and describing him as having brown hair and brown eyes.

He was to remain assigned to the depot of his regiment for several months and, no doubt because of his level of education, received his brigadier stripes on March 19, 1915. It was not until May 27, 1915, that he left for the front to join the bulk of his regiment, which was then assigned as a second-line unit in the trenches of Champagne – several detachments took part in the attack on Champagne in September 1915, before finding themselves assigned to a quiet sector of the front near Montbéliard, near Belfort. The attack on Verdun sent the 9th Hussars to the sector where they were only used to police the rear of the lines. Throughout this period, Brigadier de Boigne served in the 2nd squadron of the regiment, more concerned with fighting boredom than the enemy, if the punishment register is to be believed, where he was given nearly 40 days' accumulated detention for various acts of negligence in garrison life or acts of bad temper towards his superiors.

Young pilot

He then volunteered to join the aeronautical service and was accepted as a student pilot on June 13, 1916. He went to the Dijon flying school from where he graduated with his military license (n°4340) on August 25, 1916. Two days later, he went to the Avord training school, then did a shooting course at the Cazaux gunnery school from October 3 to 26, 1916, when he finished his training at the Pau fighter school.

After a period at the Groupement des Divisions d'Entrainement, he was assigned to escadrille N 82 on January 2, 1917. The unit was stationed at Fontaine airfield near Belfort, in a sector of the front that was reputed to be rather calm. He remained there until April 1917, when the unit was assigned to the Bonnemaison (Lhéry) airfield in Champagne to take part in the air offensive following the attack on the Chemin des Dames, where the aerial battles took on a whole new dimension.

It was there that Brigadier François de Boigne obtained his first aerial victory at the controls of his Nieuport, on May 3, 1917, by shooting down an enemy aircraft with his squadron leader Captain Echard, which fell into the French lines, earning him a citation in the Army Order on May 25 and a promotion to the rank of maréchal des logis on the following June 26. The

Above: Portrait of François de Boigne, dressed as an aviator.

N 82 began to receive SPAD VII and François de Boigne was given n°1405, but the unit was still equipped with Nieuports which were proving to be obsolete. On November 4, 1917, escadrille N 82 coming back to Nieuport received the order to fly for new skies where our young pilot distinguished himself.

Success in Italy

Following the defeat at Caporetto, the Italian army began a chaotic retreat towards the west and was only able to recover along the Piave River, at the cost of abandoning a large part of its heavy equipment, including half of its aircraft – its strength

Above & Above Right: Portrait of François de Boigne, dressed as an aviator.

dropping to 200 aircraft online.

The N 82 squadron was then sent to the other side of the Alps to rescue the Italian ally, along with several other corps squadrons, and another fighter squadron, the SPA 69. The Nieuport 27s of the N 82 squadron were recognized by the head of the French air force in Italy as good aircraft, inferior to the German planes. But the enemy facing the French on this front was the Austrian air force, whose equipment was far from being equal to that of its German ally, which allowed the French pilots of N 82 to face the enemy on an equal footing.

Installed in Verona on November 4, 1917, the N 82 had more to risk from the winter climate and the configuration of the front, made of mountains, than from the action of the enemy aviation. Most of the losses came from pilots crashing in the wilderness following a mechanical breakdown, and having no flat ground to land on... For maréchal des logis François de Boigne, the skies over Italy were synonymous with success. He was promoted to the rank of second lieutenant on a temporary basis on January 26, 1918, and the following month, on February 24, 1918, he won a "double" by shooting down two Austrian aircraft obtained in collaboration with Captain Echard – one falling on Monte Paole, the other north of Possagno. This new exploit earned him a citation in the army order on March 28 and the bronze medal for Italian military valor. On March 23, 1918, he again claimed a plane that fell on the town of Montebello, but which was not certified, and then, on March 25, 1918, a two-seater fell under his fire north of Montebelluna, which was certified in collaboration with Sergeant Bolland, his fourth victory, as stated in his new citation in the army order dated May 21, 1918.

With his three sure victories and one probable shot down on the Italian front, he was the best French fighter pilot of the sector. His successes were hailed by Major Houdemon, head of the French military aeronautical mission, who noted the impetuous side of his 21-year-old pilot, whom he also considered *"very young in character and not very serious."*

Aviation ace

François de Boigne Victory List							
N°	Date	Unit	Plane piloted	Plane claimed	Place	Notes	Confirmation
1	3-May-17	N 82	SPAD VII n°1405	Enemy plane		With Cpt Echard	Citation OA 25-5-1917
2	24-Feb-18	N 82		Enemy plane	Grappa / Monte Palone	With Cpt Echard	Citation OA 28-03-1918 - JORF 23-05-1918 p.4510 "2nd and 3rd victories"
3	24-Feb-18	N 82		Enemy plane	Grappa / N. Possagno	With Cpt Echard	Citation OA 28-03-1918 - JORF 23-05-1918 p.4510 "2nd and 3rd victories"
-	23-Mar-18	N 82		Enemy plane	Montello		
4	25-Mar-18	N 82		Two-seater	N. Montebelluna	With Sgt Boland	Citation OA 21-05-1918 - JORF 01-07-1918 p.5717 "4th victory"
-	21-May-18			Two-seater			
5	27-May-18	N 82		Fighter	Craonne	With Adj Rimbaud and MdL Faure	Citation OA 16-06-1918 -JORF 24-07-1918 p.6380 "5th victory"
6	9-Oct-18	SPA 69	SPAD	Two-seater	German lines	With Adj Franc	Citation OA (LH) - JORF 08-04-1919 p.3691 p.3691 "6th victory"
Note: German casualties column deleted because there were no entires.							

After returning to France on April 10, 1918, he immediately left for the front in the Oise region with N 82, re-equipped with SPAD fighters, to try to contain the German offensive launched in the area in the hope of achieving a decisive victory before the arrival of American troops.

It was there that he achieved his last successes: he shot down a fighter over Craonne on May 27, 1918, in collaboration with two crew members, which earned him a new citation on June 16, 1918.. Transferred to the SPA 69 squadron on September 16, 1918, he finished the war in Lorraine when he obtained his last victory on October 9, also in collaboration with a wingman.

With a final score of 6 certified victories and two probable victories, he was one of the aviation aces but remained totally anonymous – the honor of appearing in the military press release being reserved in 1918 for pilots with more than 10 successes.

Staying in the squadron for a few more months, he received the Knight's Cross of the Legion of Honor on April 8, 1919, which rewarded his 6th victory. He was then transferred to the 1st Aviation Group of Dijon on May 16, 1919, before being put on indefinite leave two months later, on July 17, 1919, and returned to civilian life.

After the war

He retired to the family castle of Buisson-Rond in Chambéry and had to think about his reconversion to civilian life. To do so, he followed the path of his distant ancestor who had gone on an adventure in India: in 1926 he became a flight instructor in Shantoung, China, where he opened an aviation school at the request of the local warlord and where he added nearly 400 hours to his flight log.

Returning to France after 1932, he found a job in the sales department of the Bugatti factory in Molsheim, where he became close to the management of the company, dating the daughter of the famous Italian manufacturer, Miss Lidia Bugatti, whom he married on March 11, 1940 in the presence of numerous personalities.

When the Second World War broke out, he was mobilized again with the rank of lieutenant: having completely neglected his training periods in the reserves, he was promoted to this rank in 1924 and received no further advancement. As soon as he was mobilized, he was given a special assignment to the Bugatti factory in Molsheim, which was then transferred to Bordeaux, at the request of his management. His special assignment was not renewed and on May 16, 1940, he was assigned to the Tours air base, to the air battalion n°109.

Demobilized after the debacle, he retired to Chambéry at Buisson-Rond and no details are known about his actions during the occupation, as he was not a member of the Resistance or the collaboration. The Air Force discharged him from its ranks on August 18, 1948, once he had reached the age limit for his rank. Active in the social life and frequenting the Kennedy brothers in the 1950s, he died on August 23, 1970 at the American hospital in Neuilly and is buried in the Barberaz cemetery in Chambéry.

Adolphe Du Bois d'Aische

Name: du Bois d'Aische (Count).
First names: Adolphe, Aloïs de Gonzague, Marie, Hubert, Gislain name).
Born: 17-03-1874 in Brussels (Belgium).
Father's profession: Annuitant.
Military pilot brevet: 25-06-1918 (n°14218).
Units: F 71.
Army communiqué: No.
Total score: 6 victories, 6 probables.
Died: 07-10-1958 in Saint-Quay-Portrieux (Côtes d'Armor).

Above: Portrait of Adolphe du Bois d'Aische.

A son of Belgian aristocracy

Adolphe Aloïs de Gonzague Marie Hubert Ghislain du Bois d'Aische was born in Brussels on March 17, 1874, from a large family of Walloon nobility whose origins date back at least to the 15th century. He was the son of Count Adrien du Bois d'Aische and Baroness Marie Thérèse de Fierlant, and as the eldest of the family's six children, he was entitled to his father's title of Count. However, this young man who lacked nothing soon proved to be a particularly turbulent child. He had many disagreements with his father, who wanted him to study law, while he only dreamed of becoming an engineer. When the Count planned to marry his son to a handsome party, the rupture was consummated and Adolphe du Bois d'Aische fled to the Belgian Congo at the turn of the century and began several years of adventures of which little is known except that he enlisted in the French Foreign Legion at the age of 30, on September 4, 1904, and served in the first foreign regiment. There he met many German legionnaires and acquired a good command of the language of Goethe. Obtaining French nationality by right, he returned to France and met a Breton woman two years his junior named Constance Catesson, who had two daughters from a first marriage. He married her in 1914 and the family settled in Paris in the 15th arrondissement; he then worked as an engineer.

But like millions of Europeans, the war took him away from his home and he was drafted into the army. He was then 40 years old and participated in the first campaigns in three different infantry regiments (the 4th, 131st, and 89th), being named corporal and then obtaining a citation in the order of the 131st regiment on February 26, 1915: "*Nice attitude in front of enemy fire. Valuable auxiliary for his company commander*". Wounded in combat, he was rested and transferred to the air force on July 5, 1915 as a mechanic, being assigned to the Avord base. He became an "R.A.T." (Territorial Army Reservist), he could expect to finish the rest of the war in a second line unit far from the fighting, given his age and his service record. But du Bois d'Aische volunteered to serve in

Above: Adolphe du Bois d'Aische as a teenager during the years 1880s in Belgium. A turbulent child, his disputes with his father, the Count du Bois d'Aische, were frequent.

Above: Adolphe du Bois d'Aische (right) in a boat around year 1900. He flew his family to Belgian Congo and enlisted in the French foreign legion, adopting French citizenship after his service.

a combat unit, and after training at the schools of Pau and Cazaux, which he completed at the beginning of February 1916, he left on April 6, 1916 as a machine-gunner corporal in the F 71 squadron, which was stationed in Commercy opposite the Saint Mihiel salient.

In the battle of Verdun

France was in the middle of the Battle of Verdun, which resulted in an unprecedented aerial battle in which the entire French fighter force, flying the small Nieuport 11 and 16, finally won air superiority over the German Fokker planes, with aces such as Navarre and Nungesser in the foreground. But the Germans did not remain inactive and made life difficult for the French observation planes, many of which were shot down... The F 71 was lucky so far, with 3 damaged planes coming back with wounded on board, in return a victory was obtained in the German lines by Lieutenant Pierry in February. The squadron flew Farman F40s, which were not particularly appreciated by the crews because of their slowness, fragility, and lack of rearward defense due to their sluggish engines. Du Bois d'Aische took part in his first missions and had an encounter on May 22, 1916:

"Aboard an old Farman, with Lieutenant Weiss, during a photographic mission in the Varvinay region, I have for the first time the joy of seeing Boche. There were three of them. My pilot, while maneuvering, found a way to deceive them by pretending to flee. Soon one of the three Albatrosses, having fallen into the trap, abandoned his comrades to give chase. It was what we were waiting for. As soon as we saw him far enough away from the others, in a sudden turn we faced him and at fifty meters I had the pleasure to start a short fight and to see my opponent crashing to the ground. It was my first try. My pilot was cited in the army order. "

The victory was recorded in the army newsletter and the two men were celebrated at the squadron. It was then decided to form an air defense section: some observation squadrons, such as Captain Vuillemin's C 11, had a few fighters in their equipment to protect their aircraft. But not in the F 71, where the escort aircraft were simply good old Maurice Farmans. Nevertheless, Du Bois d'Aische remained enthusiastic:

"For us machine gunners, it was the prospect of Homeric battles; it was joy and good times, because as soon as Boche planes were reported heading towards our lines, in five or six minutes we took to the air, with

Above: Portrait taken in 1905, probably in France where he worked as an engineer in Paris.

four aircraft grouped together, in search of the Boche. The beginnings were unforgettable moments. The comrades, determined to dare everything, were waiting, like me, for the opportunity to fight. Many were our disappointments. On June 10 four Boche, seen, could not be reached; on the 14th, another one escaped us."

On June 16, the Farmans of F 71 came across a formation of a dozen German aircraft, which they were able to intercept by cutting off their retreat. A melee ensued, but du Bois d'Aische was unable to fire because the ejector of his machine gun broke. He then grabbed the musket he was carrying and fired at the enemy – one German fell to his death and left the fray. He did not get the nod, but another opportunity soon presented itself.

"On June 21, returning from a successful daylight bombardment in the Conflans-Buxerulles area, at point-blank range, I had the satisfaction of riddling a Boche who was attacking us and shooting him down in flames in his lines, but too far away for the infantry to have seen him. This last again is not approved."

The German archives tell us, however, of the loss of a two-seater of FA 3b, shot down at Vigneulles (less than 10 km north of Buxerulles), resulting in the death of its crew, Sergeant Alois Hosp and Lt. Erich Deliser. But nothing is less certain, as 2nd Lieutenant Jean Chaput, an ace of N 57, claimed his 6th victory in this sector. In any case, Du Bois d'Aische was rewarded by being appointed sergeant in the following weeks. Then enemy air activity became scarce and encounters diminished, although the flak was still intense: *"Twice I had my pilot wounded by shrapnel; we often returned with our aircraft seriously damaged, but no Boche."*

Above: Corporal du Bois d'Aische (2nd left on the ground) at the Cazaux gunnery school in late 1915.

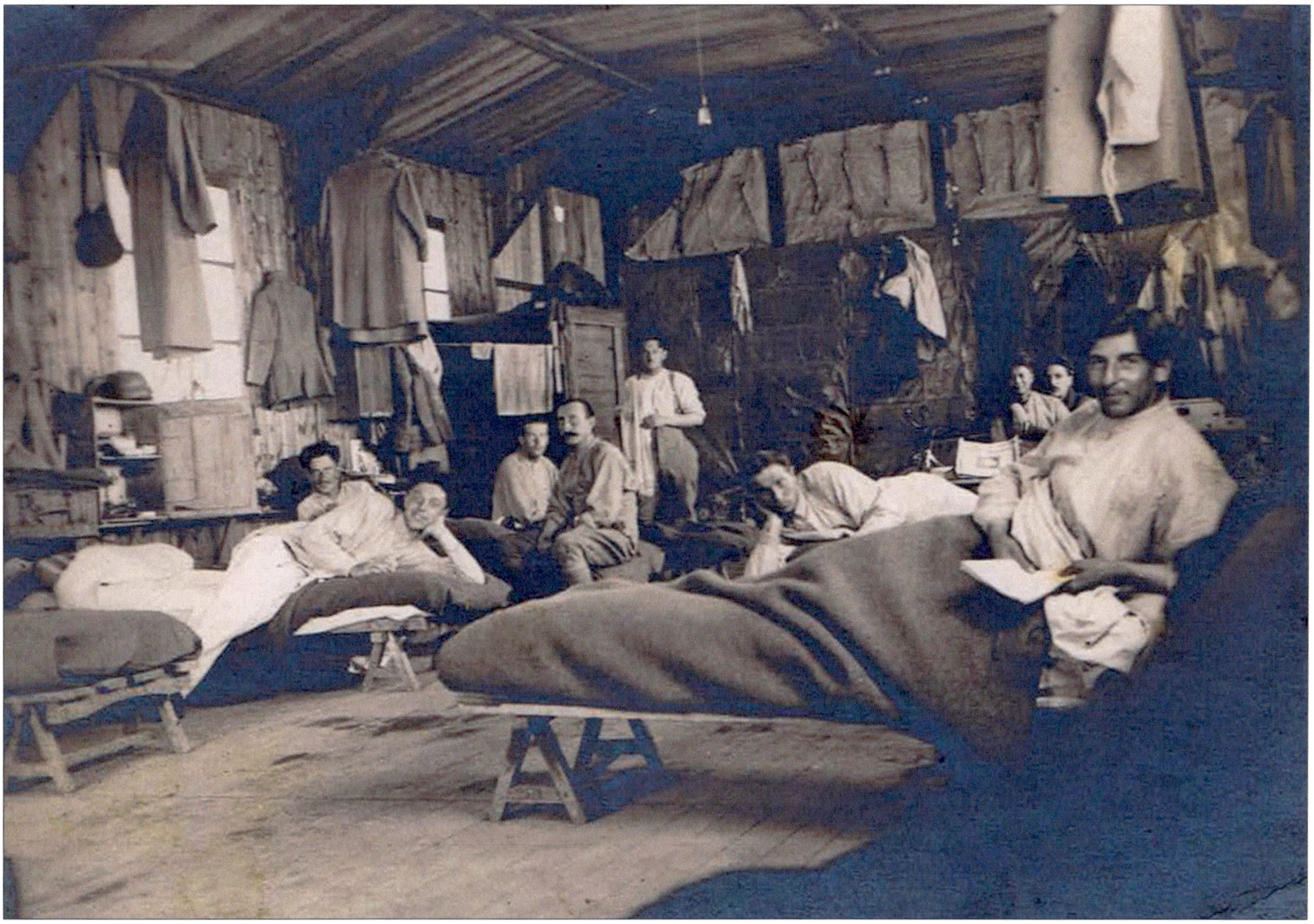

Above: In his dormitory in Cazaux.

The Chemin des Dames offensive

The F 71 squadron was then moved in September 1916 near Nancy, then briefly at the end of December to the Somme, before settling on January 23, 1917 at the Sainte Ménehould airfield, at the western end of the Verdun sector, where the main fighting had now ended. The action now took place to the west, on the Craonne plateau, where General Nivelle launched the Chemin des Dames offensive from April to June, which ended in a bitter failure and led to mutinies among the troops. Three or four Salmson-Moineau SM 1s, ungainly aircraft with a radial engine fixed along the fuselage, driving two propellers by means of complex transmission shafts, were delivered to the squadron as a complement to its Farman aircraft, which were now completely outdated. The squadron suffered badly in April when four crews were wounded; in May it was a Salmson-Moineau that was shot down. Du Bois d'Aische was lucky enough not to come across any enemy aircraft before the arrival of the Sopwith 1 A2s to which he was assigned at the end of May.

"*From then on, I was lucky. On June 3, with pilot Fontaine (now dead), during a hike along the lines, five Boche fighters appeared in the sky. My heart was pounding; I showed them to my pilot. I hope they don't turn back. But no, they are coming. They are above us, seemingly looking for the point to attack. I regained my composure, and my pilot, who had eyes only for me, executed all the maneuvers agreed upon in advance. The first Boche spiked and, shot at less than thirty meters, continued his vertiginous descent to the ground. The second one follows the same path and, clearly hit, also crashes into a wood. The last three seem to hesitate. My pilot and I were too excited to think of anything else but attacking them. A few well-aimed bursts of machine-gun fire caused one of the adversaries to land helplessly, while the last two, much faster than our Sop, sought their salvation in flight. We cannot think of joining them and we return our mission completed.* "

Only one victory was recorded for him north of Laffaux, near Soissons, west of the Chemin des Dames; he was also cited in the army order on June 17, 1917. The German archives are silent on the loss of hunters on this date. On the other hand, they tell us about the loss of a two-seater Schsta 26 shot down at Craonnelle, 10 km from Laffaux, resulting in the death of its crew, Vfw Friedrich Beisecker (pilot) and Flg. Pirmin Bücklein (observer), but there were other candidates for this victory, including Corporal Guérin of the N15 (claiming his second victory at about the same place) and the French ace Georges Guynemer, who claimed an unconfirmed two-seater at Craonne, much closer to where the German fell.

The fighting continued and on June 28, 1917, du Bois d'Aische almost got shot down while flying an AR with adjudant René Pons as pilot. The two men spotted a large

Above: Maurice Farman F 40 of escadrille F 71 during spring 1916.

German two-seater and managed to approach it. It accepted the fight and its gunner proved to be particularly skillful, managing to cut the warp controls of the AR at the first burst, which went into a spin. Fortunately, Pons managed to regain control of it, and, using all his skill as a pilot, managed to regain his ground. He received a well-deserved commendation... On July 24, 1917, Adolphe du Bois d'Aische won two certified victories during the same sortie, his first official "double", during a very difficult fight where he saw death even closer at Cernay en Dormois.

Above: Ceremony in front of the Farman F40s of escadrille F 71 during year 1916.

The AR 71 squadron left that day on a photographic mission 15 km behind enemy lines. An AR, mounted by the crew MdL Clément (pilot) and Lt Reibell (observer), was in charge of taking the pictures. It was escorted by 3 other ARs which adopted a defensive diamond formation, with on its left the crew Sgt Bétis / Du Bois d'Aische, on the right the crew Sgt Boulaire / Cprl Lahémade, and behind the AR mounted by the crew Adj Pons / Sgt Couture, the most dangerous place. As they crossed the lines, the flak reacted weakly, a sign that German hunting was in the area. It did not take long for them to appear: 7 Albatros and Roland fighters fell on them from behind, targeting the aircraft to the rear and left of the formation. The gunners opened fire and the sky was strewn with tracer bullets. Sgt Couture hit an enemy fighter that went into a dive, but he was immediately hit by two Germans who riddled his AR and wounded the pilot Pons. The latter had no other resource than to engage in a desperate dive pursued by an Albatros. The AR of Sgt Boulaire dove in turn and shot at the pursuer, making it clear and allowing Pons to land in his lines. At the same time, Du Bois d'Aische, caught by two fighters, used his Lewis gun successfully, hitting one that was seen crashing to the ground, and a second that spiked, breaking up the fight. But in the firefight, his pilot Betis is shot in the stomach and the plane goes into a spin. "*Falling from spin to spin into the third line trenches, I considered myself coldly doomed.*" Fortunately, Bétis managed to regain control of his plane in extremis, and to crash-land his plane, which broke up on contact with the ground. Rescued by French soldiers who witnessed the fight, Du Bois d'Aische came out of the wreckage more or less intact, but Sergeant Henri Bétis was not so lucky as he died of his wounds three days later in a field hospital in Sainte Ménehould. The German archives tell us that Leutnant Oskar Dänkert, of Jasta 9, based at Leffincourt, west of Verdun, in the sector of AR 71, was the only French squadron to make a claim that day. Its four crews who participated in the battle were cited in the Army Order on July 31, 1917 – only two ARs, those piloted by Clément and Boulaire, were able to return to their base.

During the summer, the Letord reconnaissance aircraft were delivered, twin-engine aircraft equipped with a firing position in the front and another in the rear. It was on this aircraft that the ace closed his record on September 22, 1917:

"*On September 22, on a Letord aircraft, with Fontaine as pilot, we were leaving to take pictures behind the Boche lines when, on the way, three Albatros fighters came at us. The pleasure is always the same but, I don't know why, I don't even fear that they will move away, they shoot from a distance, so they have good intentions. I make my arrangements and wait for them. The first one arrives soon at a good distance and between thirty and forty meters three or four bullets send him in flames to the ground. The others did not insist and fled. We continued quietly when we saw a large two-seater plane that seemed to be scouting. We charged at it and the fight was so close that we could clearly see every detail on board. So it was not long and after a few exchanged bullets it fell and we saw it crash on the ground. This time luck favored me because both planes were certified and gave rise to a citation to the armies on October 12, 1917. Since then, Fontaine died while trying an aircraft, and luck did not want me to have a single combat, especially since the flights are rare because the weather is constantly foggy or rainy.* "

Above: Corporal Adolphe du Bois d'Aische with his dog, at escadrille F 71 early 1916.

No German fighter pilots are recorded as being killed or wounded that day; however, two two-seaters fell within 50 kilometers of Sainte Ménehould, within flight range of the Letord but the lack of a place of claim prevents us from knowing more. Du Bois d'Aische officially became an ace with his six certified victories and was awarded the military medal by General Hely d'Oissel, commander of the 8th Army Corps, during a ceremony at Villers Daucourt in October.

He then asked for a special favor: at the age of 43, he wanted to become a fighter pilot. Although schools were not allowed to accept candidates over 30 years old, he was granted this privilege and, after a well-deserved three-week leave in October 1917, he joined the Cazaux school in December and then Istres, where he was licensed as a military pilot (n°14 218) on June 25, 1918. After training in Pau and Voves, he joined the GDE on September 18, 1918, but did not have time to be assigned to an operational unit. Wounded in an accident on SPAD VII on October 3rd, he was transferred on December 10th 1918 to the Chartres school.

Employed after his demobilization in the civil aviation, he will exercise various functions on airfields during the inter-war period, working in Ajaccio until 1922 as the hydrobase director, and then at Le Bourget airfield, where he has important responsibilities before being dismissed following

Facing Page, Above: In the cockpit, Adolphe du Bois d'Aische is goofing around in his Salmson-Moineau escort plane. A difficult, heavy plane to fly and even more dangerous to land, the Salmon-Moineau had only quality , its two gunners positions, but was disliked by its crew and was quickly replaced by the Letord.

Facing Page Below, Above, & Below: Dorand AR 1 of sergent Bétis (left medallion) and corporal du Bois d'Aische (right), shot down by German fighters on July 24, 1917. Du Bois d'Aische was miraculously unhurt, while Bétis died from his wounds.

Above: Adolphe du Bois d'Aische and lieutenant observer Georges Laurent near the graves of sergent Bétis and sergent Couture, killed in action on July 24, 1917.

a violent dispute with his minister... When the Second World War broke out, he was a 65 year-old man retired and living in the Breton village of Saint Quay Portrieux, where his wife was born. Nevertheless, he was employed by the German administration as an interpreter and took advantage of this to warn the Resistance fighters who were about to be arrested. Unfortunately, he was discovered and imprisoned in a fortress. According to him, despite the beatings and insults, he never deigned to work! Released in 1944, his activity for the resistance would be officially recognized and would even be the subject of a file of approval by the authorities. Still living in Saint Quay Portrieux after the war with his wife, he was elected town councilor of the city and died on October 7, 1958 at the age of 84.

The tricks of the ace

What to remember about this atypical ace, the eldest of all French WW aces? During the years 1916–1917, all causes considered, escadrille F 71, which was far from being a special case, lost 11 airmen killed, 4 prisoners, and 18 wounded, some of them several times. So there was a certain factor of luck, to have passed through the Russian roulette of the flak and to have made rather limited aerial encounters. Then, the aggressiveness: he admits himself, in the letter that he addresses to the journalist Jacques Mortane, of a strong bad character, which caused him "many injustices". Pitying not to meet any German in the sky, each enemy met was systematically attacked... Then remains the technique. If the military authorities recognize him for the figure of six approved victories, his real record will always remain unknown. The German archives, incomplete, can make us deny or confirm certain claims; but how many aircraft were forced to turn back, hit by the precision of its defensive fire? Faced with a World War I fighter armed at best with two 7.7 mm weapons and with a speed superiority of barely 50 km/h, a sniper, who does not panic and manages to give instructions to the pilot so that the latter puts him in firing position, is on equal terms. Let's leave it to him to conclude:

"*When in the infinite blue sky you see a few small black dots, your heart beats, it's true, it happens often. But is it the Boche you want so badly? That's what you can only know when he comes within 3 or 400 meters of you, and many are the disappointments. Unless there are several of them, they don't attack and faster than our adjustment devices, they vanish without any hope of reaching them. When you see them at five or six, then the score changes. The fear of seeing them disappear like smoke gives way to real joy. You take your battle dispositions. You warn your pilot who, attentive to the slightest gesture, maneuvers and attracts the adversary who is stupidly shot at thirty or forty meters. One would really*

Above: Adolphe du Bois d'Aische and lieutenant observer Georges Laurent near a Dorand AR 1, escadrille F 71, August 1917.

Above: Portrait near a Dorand AR 1, August 1917.

have to be of the last clumsiness not to shoot down such an easy prey. "

Above: The airfield of Sainte Menehould during spring 1917, housing escadrille F 71. Note the Dorand AR two-seaters on the right.

Above: Letord 1 of escadrille F 71 in 1917.

Below: Lt. Georges Vicent, unknown pilot and Sergent Adolphe du Bois d'Aische near a Letord 1 in 1917.

Above: All F 71 personnel pose in front a Letord 2 decorated with a parrot, the symbol of the unit. Adolphe du Bois d'Aische stands in the gunner position, with Lieutenant observer Georges Vincent behind him.

Above: During the month of August 1917, Sergent Adolphe du Bois d'Aische is decorated of the military medal by General Hely d'Oissel in Sainte Menehould airfield.

Above: Adolphe du Bois d'Aische in a Nieuport 21 as a student pilot in 1918… Aged 44!

Farman F 40 of escadrille F 71 during year 1916, the first type of plane where Adolphe du Bois d'Aische made his first operational flights as a gunner.

Above: Count Adolphe du Bois d'Aische in his eighties during the 1950's in his home in Britanny with his family.

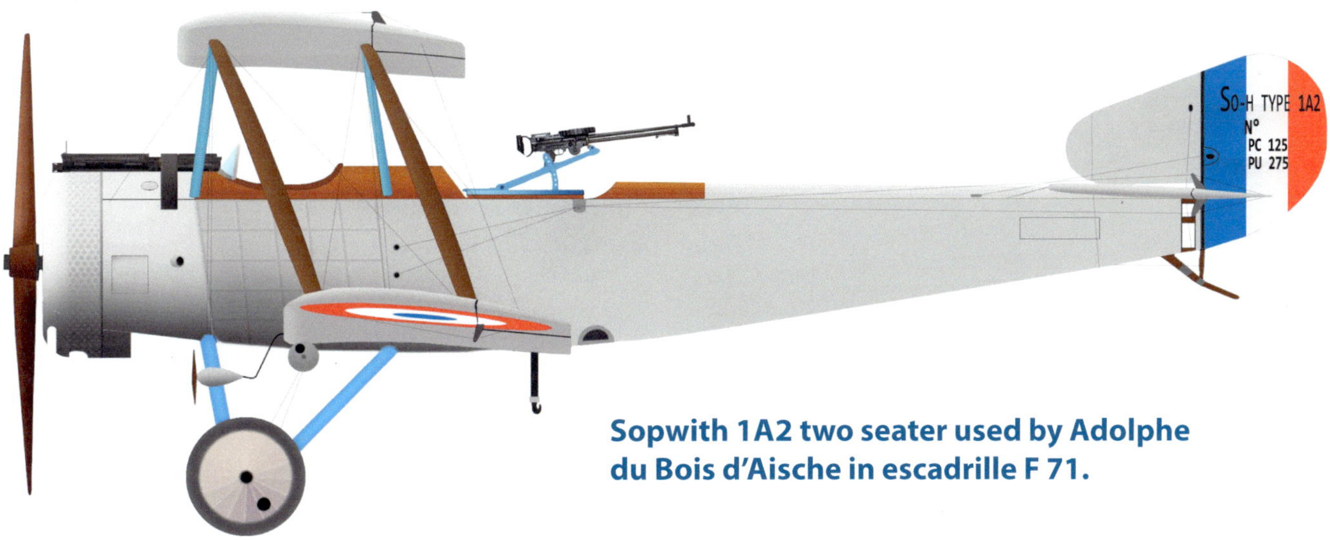

Sopwith 1A2 two seater used by Adolphe du Bois d'Aische in escadrille F 71.

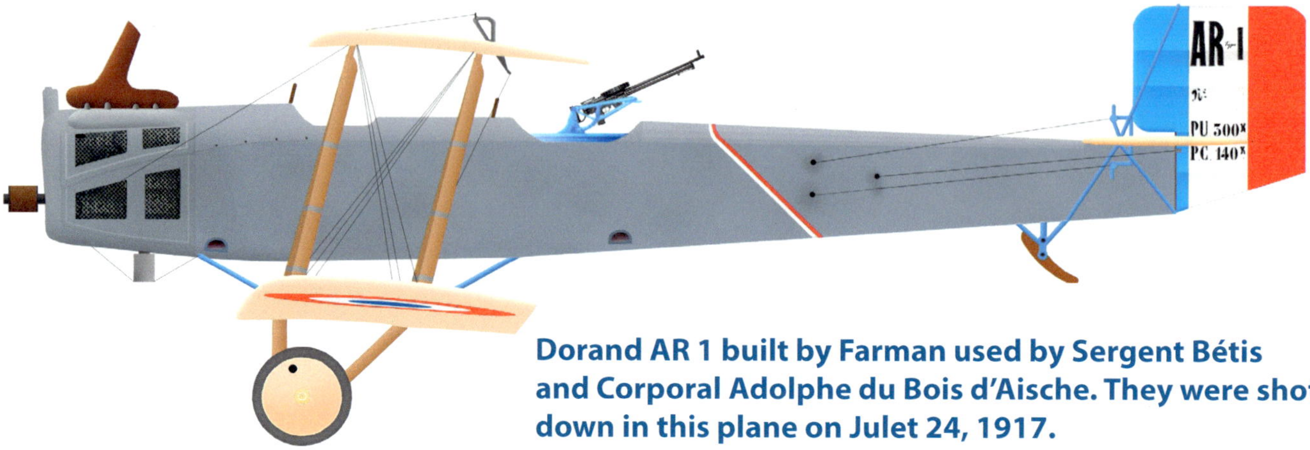

Dorand AR 1 built by Farman used by Sergent Bétis and Corporal Adolphe du Bois d'Aische. They were shot down in this plane on Julet 24, 1917.

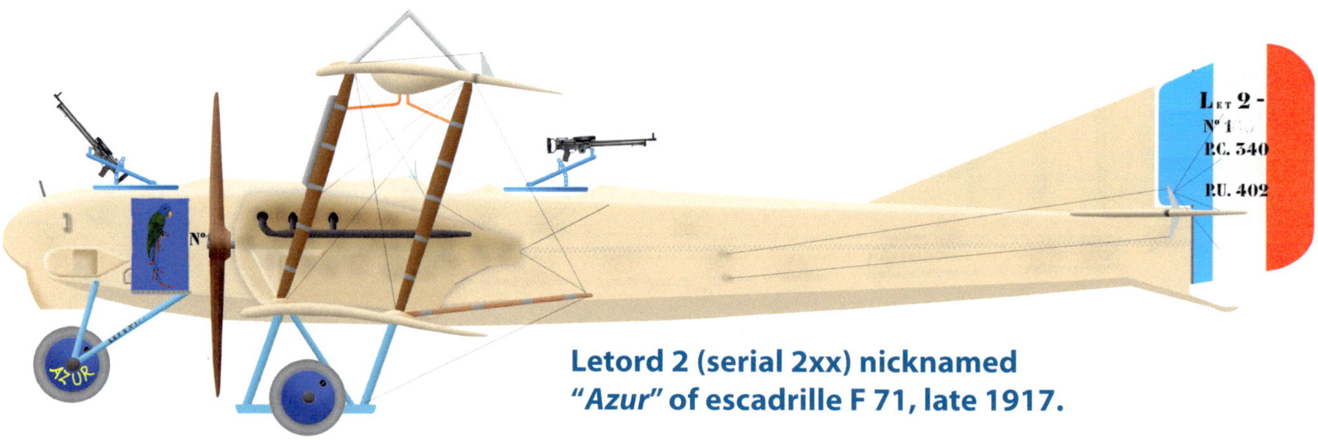

Letord 2 (serial 2xx) nicknamed "*Azur*" of escadrille F 71, late 1917.

Adolphe du Bois d'Aische Victory List							
N°	Date	Unit	Plane piloted	Plane claimed	Place	Notes	Confirmation
1	22-May-16	F 71	Farman F 40	Albatros	Varvinay	With Lt Pierre Weiss (pilot)	
P1	16-Jun-16	F 71	Farman F 40	Two-seater	St Mihiel	With unknown pilot	
P2	21-Jun-16	F 71	Farman F 40	Enemy plane	Buxerulles	With unknown pilot	
2	3-Jun-17	F 71	Sopwith 1A2	Fighter	N. Laffaux	With Brig Jacques Fontaine (pilot)	
P3	3-Jun-17	F 71	Sopwith 1A2	Fighter	N. Laffaux	With Brig Jacques Fontaine (pilot)	
P4	3-Jun-17	F 71	Sopwith 1A2	Fighter	N. Laffaux	With Brig Jacques Fontaine (pilot)	
3	24-Jul-17	F 71	Dorand AR	Fighter	Cernay-en-Dormois	With Cprl Henri Bétis (pilot)	Citation OA 31-07-1917 - JORF 15-11-1917 p.9121 "shot down two enemy planes"
4	24-Jul-17	F 71	Dorand AR	Fighter	Cernay-en-Dormois	With Cprl Henri Bétis (pilot)	Citation OA 31-07-1917 - JORF 15-11-1917 p.9121 "shot down two enemy planes"
5	22-Sep-17	F 71	Letord	Albatros D	N. Bois de L'Echelle	With Brig Jacques Fontaine (pilot) and Lt Marcel Meunier (front gunner)	Citation OA 13-10-1917 - JORF 11-12-1917 p.10043 "shot down one fighter and one two-seater, raising his victory number to six"
P5	22-Sep-17	F 71	Letord	Enemy plane	Vouziers / Challerange	With Brig Jacques Fontaine (pilot) and Lt Marcel Meunier (front gunner)	
P6	22-Sep-17	F 71	Letord	Enemy plane	Vouziers / Challerange	With Brig Jacques Fontaine (pilot) and Lt Marcel Meunier (front gunner)	
6	22-Sep-17	F 71	Letord	Two-seater	Bois Sugnon	With Brig Jacques Fontaine (pilot) and Lt Marcel Meunier (front gunner)	Citation OA 13-10-1917 - JORF 11-12-1917 p.10043 "shot down one fighter and one two-seater, raising his victory number to six"
German casualties column has been deleted because there is no information available.							

Honoré De Bonald

Name: de Bonald (Viscount).
First names: Honoré, Marie Léon, Joseph, Guillaume.
Born: 13-08-1894 in Valady (Aveyron).
Father's profession: Annuitant.
Military pilot brevet: 30-12-1915 (n°2220).
Units: N 69.
Army communiqué: No.
Total score: 5 victories, 1 probable.
Died: 07-11-1977 (place unknown).

Above: Honoré de Bonald, disfigured by a wound in the mouth received in 1914.

A young man from the aristocracy

When he was born in Valady (Aveyron) on August 13, 1894, the young Honoré, Marie Léon, Joseph, Guillaume, de Bonald was particularly well born. Son of Viscount Joseph de Bonald, he came from an old French aristocratic family from Millau, where his ancestors had held the office of judge-bailiff for seven generations. Among its illustrious members, let us quote his great-great-great-grandfather Louis de Bonald (1754–1840), theorist of the counter-revolution, minister under the Restoration and member of the French Academy; as well as Victor de Bonald (1814–1897), mayor of Montpellier and then of Millau, and monarchist congressman of Aveyron from 1871 to 1876.

The young man spent his adolescence in Toulouse with his two older sisters Marie Madeleine and Marguerite Marie, the family settling at 13 rue Velane. Probably attending college, he benefited from a careful education and acquired a great general culture from his father, who devoted himself to historical studies and collaborated with the historical studies group of Toulouse.

Voluntary enlistment

The young man was destined for a military career and, before his 19th birthday, he signed a voluntary enlistment in the army on April 19, 1913 for a period of 4 years at the Toulouse City Hall. He chose to be incorporated into the aristocratic arm, the cavalry, and found himself incorporated a few days later into the 12th regiment of hussars stationed at Gray in Burgundy.

He was quickly promoted to brigadier on October 20, 1914, and then to maréchal des logis on July 2, 1914, one month before the outbreak of the First World War. At the outbreak of hostilities, the regiment was engaged in the French offensive in Alsace and took part in the first fighting in the Vosges forest. On August 23, 1914, maréchal des logis de Bonald was wounded by a bullet that went through his mouth. Evacuated, he was treated in a hospital in Lyon and left disfigured for life, with a deviation of the mouth that made it difficult for him to open his mouth and caused him chronic gastrointestinal problems.

Nevertheless, he returned to combat and rejoined his unit before the end of 1914, being posted to the Somme front and to Champagne. However, noting the uselessness of the cavalry in this trench warfare, he asked to be transferred to the air force and was accepted as a student pilot when he went to the Pau school on September 16, 1915, where he was appointed to the rank of second lieutenant on a temporary basis on December 26, 1915, and where he obtained his military pilot's license (n°2220) four days later. On January 16, 1916, he went

Above: Honoré de Bonald in his SPAD VII from escadrille N 69.

to the Avord school where he perfected his skills in flying the Nieuport fighter. On April 4, 1916, he joined the Bar-le-Duc group in the middle of the Verdun battle and was assigned two days later to the N 69 squadron.

Fighter pilot

He flew his first missions in the Verdun sector with a Nieuport 11 or 16 fighter, before his unit migrated to the Somme front and settled at Grivesnes on July 19, 1916. It was there that he claimed his first success against an aircraft that was shot down on August 1, 1916, but for which he did not obtain certification. He had to wait until November 1st 1916 to obtain his first official victory against a plane shot down east of Ablaincourt in collaboration with Corporal Boy.

The N 69 then left the Somme to settle in Champagne in March 1917, while the French offensive on the Chemin des Dames was being prepared. It was there that De Bonald, promoted to lieutenant on December 16, 1916, achieved most of his successes, shooting down an aircraft on March 28, then a fighter on Grignicourt on May 4, and a DFW C with two wingmen on June 3, 1917, north of Muison. Unfortunately, on June 14, 1917, he was seriously wounded in aerial combat by a bullet that went through his thigh.

Last victory on the Italian front

Evacuated the next day, he had to spend several months in hospital and convalescence, from which he was to suffer from a lifelong difficulty in walking. He returned to his unit on November 12, 1917, when it was re-equipped with SPAD fighters and left for Italy. Indeed, attached to the 10th French Army, the SPA 69 followed it when it was sent to the rescue of the transalpine ally, whose front collapsed following the defeat of Caporetto.

Installed on the field of Porto Nuova di Verona, then San Pietro in Gù from December 17, 1917, the pilots of the SPA 69 discovered a very different sector than the French front: the winter cold was much more intense because of the mountains which also made any emergency landing very risky. But the enemy air force, composed of Austrian squadrons, proved to be much less biting. This allowed the pilots of SPA 69, led by Captain Malaviale, to attack the enemy directly on its airfields: on March 16, 1918, Lieutenant de Bonald led a patrol to attack an Austrian airfield and fired on the hangars. A two-seater rolled to take off and was immediately set on fire by de Bonald in collaboration with his teammate, Sergeant Balzac. This victory, questionable because it was obtained on the ground according to the report that was written on the attack, was

Above: Pilots of escadrilles SPA 69 visiting their comrades of escadrille N 561 in Venice (Lido), early 1918. From left to right, in the car: unknown driver, Lt de Limur (N 561), unknown, Cpt Paul Malavialle (6 victories, SPA 69), Lt Honoré de Bonald (SPA 69). Outside the car: Lt Frédéric Loiseau (N 561), S/Lt Jean Bignon (N 561).

nevertheless recognized as his 5th aerial victory by a citation in the army order – two newspapers, *La Croix* and *Le Matin*, saluted the arrival of a new ace in an insert in their inside pages.

Back in France on April 10, 1918 with his squadron, Lieutenant de Bonald fought again in the Somme and in the Aisne until the final victory, without increasing his score and without knowing the honor of the communiqué which was then reserved for pilots having obtained 10 victories. He ended the war with the Legion of Honor and the Croix de Guerre, decorated with 5 palms and corresponding to as many commendations in the army order, obtained in 400 hours of war flight over the enemy.

Career in military aviation in the 1920s

As an active soldier, he remained in his unit after the armistice and was transferred on January 11, 1919 to escadrille SPA 94, where he was put in command. He did not stay there for long, as he was assigned on January 15 to the French military mission in Finland, where he stayed for almost a year, returning to France on February 23, 1920 where he was assigned to the 2nd observation aviation regiment. On July 14, 1920, he left for North Africa where he was assigned to the Algeria-Tunisia aviation regiment, and then, on March 20, 1922, he was seconded to the 15th army corps staff as an aviation officer.

He stayed there for less than three months, before being assigned to the Istres aviation school where he was promoted to the rank of captain on June 25, 1923. His career took a new turn on October 25, 1925 when he was assigned to the 39th aviation regiment in the Levant, becoming assistant to the commander of the Damascus-Deraa aviation group. He was always highly rated in each of his assignments, and his superiors did not fail to underline his great general and technical knowledge, as well as his tact and his sense of command - the very type of the perfect officer.

Health problems

Unfortunately for him, his war wounds came back to surface. He had to be repatriated from Levant on September 22, 1927, where he returned to the school in Istres. The Marseilles commission already proposed that he be put aside with a temporary disability pension of 40%. He managed to avoid

Coat of arms of the De Bonald family.

this for a while, but his health problems were not resolved. At that time his father died in 1930, the year he got married on October 15 to a young woman of 32 years as well born as him, Miss Anne Charlotte Alix Henriette de Lastic, daughter of a marquis. The marriage of the new Viscount de Bonald, which takes place in Parentignat, in the Puy de Dôme, is announced in the aristocratic social newspaper "*Le Figaro*", the air force general Massenet de Marancourt being one of the witnesses of the groom.

Honoré de Bonald had time to devote to his wife: on August 1, 1931, he was retired due to his state of health with the rank of commander. He then retired to the Château de Monna, near Millau, the ancestral home of the de Bonald family. Four children were born from his marriage, including his son Jean on July 23, 1932.

Now a reserve officer, de Bonald was appointed

SPAD VII nicknamed "*Willy*" of Lt Honoré de Bonald, SPA 69, 1917.

commander of the Montpellier regional air force. When the Second World War broke out, he was mobilized again on September 3, 1939 and sent to an air force school. However, his health was quickly judged to be "precarious and alarming". Sent home on December 21, 1939, he was permanently disbarred on May 3, 1940 after being diagnosed with an intestinal obstruction.

He retired to his home where he was treated and did not take part in the Second World War or in the aviation world. He died on November 7, 1977 at the age of 83.

Honoré De Boland Victory List

N°	Date	Unit	Plane piloted	Plane claimed	Place	Notes	German casualties	Confirmation
P1	1-Aug-16	N 69		Enemy airplane				
1	1-Nov-16	N 69		Enemy airplane	E. Ablaincourt	With Cprl Boy		
2	28-Mar-17	N 69	SPAD	Enemy airplane				
3	4-May-17	N 69	SPAD	Fighter	Guignicourt		Jasta 21, Vfw Hans Brikmann and Vfw Eduart Horn killed	
4	3-Jun-17	N 69	SPAD	DFW C	N. Muison	With MdL Soulier et Sgt Chapelle		Citation OA (LH) 16-06-1917 - JORF 17-08-1917 p.6461 "4th enemy plane"
5	16-Mar-18	SPA 69	SPAD	Austrian airplane	Italy	With Sgt Balzac. Airplane taking off.		Citation OA 28-03-1918 and 10-12-1918 "5 airplanes shot down"

Above: Nieuport 27 of escadrille N 561 in Venice, 1918. Most fighters of this unit had colorful personnal decorations.

Above: SPAD XIII of escadrille N 69 in Italy.

Robert De Bonnefoy

Name: de Bonnefoy (Count).
First names: Joseph, Denis, Bernard, Robert (usual name).
Born: 29-10-1892 in Saint-Jean-le-vieux (Ain).
Father's profession: Army officer (lieutenant).
Military pilot brevet: 31-05-1915 (n°1016).
Units: VB 101, N 68, N 65, N 84, BR 131, SPA 23.
Army communiqué: 13-11-1916 for 5 victories.
Total score: 6 victories, 6 probables.
Died: 27-09-1946 in Saint-Jean-le-Vieux (Ain) (drown in Ain river).

Above: Robert De Bonnefoy portrait.

A child with a strong temper

Joseph, Denis, Bernard, Robert de Bonnefoy was born on October 29, 1894 in St Jean le Vieux, in the Ain region, into an aristocratic family where his father was a viscount and owned a castle in a place called Hauterive. Taking Robert as his first name, he was born in his grandmother's house, because his parents' declared residence was in the commune of Ste Menehould, in the Marne, where his father served as a lieutenant in the 8th cuirassiers regiment.

But if his family had no financial worries, the childhood of the young boy, who had a sister two years older than him, was not without problems. The authoritarianism of his father did not go down well with Robert, who declared in adulthood about his father: "*He has a brain as big as a thimble, he is a brute and I have absolutely nothing of him. He used to beat me as a child, and it was as a result of this that I had an asthma attack because I was so angry.*"

He indeed suffered from asthma attacks at the age of 7 or 8 and quickly became a rebellious child. When he entered junior high school, he proved to be boisterous and rowdy. Like many boys of his generation, he was fascinated by the emerging aviation industry and at the age of 17 he decided to become a pilot. Faced with his family's refusal, he said he would not leave his room until he was satisfied...

Voluntary enlisted in the cavalry

He finally reached his goal: after passing his baccalaureate in science, he obtained his civil pilot's license on January 10, 1913 at the Deperdussin school, having no doubt paid for the courses with his savings. He had to think about working: he decided to enlist for 3 years in the army on March 11, 1913 in the 5th cuirassiers regiment.

He was 1m71 tall and weighed 64 kg. He declared at the time of his incorporation that he wanted to become a cavalry officer and that he wanted to prepare for the Saumur school, or a non-commissioned officer in the military aviation. But for the time being he had to serve as a simple 2nd class cavalryman in his regiment. If he was arrested for several days for negligence, these were probably minor because he was appointed to the rank of brigadier on February 14, 1914.

He was still at this rank when the war broke out in August 1914 and went to fight with his unit in Belgium in the first weeks of the conflict. Transferred to the 8th cuirassiers regiment on October 4, 1914, after the battle of the Marne, he soon moved into aviation: the flying schools having been closed at the beginning of the conflict by General Bernard, who was in charge of aeronautics, they were hastily reopened by his successor, General Hirschauer, who immediately recruited soldiers who had their civilian pilot's license.

Above: Hauterive, the family castle of De Bonnefoy's family in Saint-Jean-le-vieux (Ain).

Passage into military aviation

Robert de Bonnefoy was one of them and was assigned on December 11, 1914 to the 1st aviation group, joining the Pau flying school on January 10, 1915 and the Avord school on March 1, 1915. His career as a military aviator almost ended before it even began, because Lieutenant Henri Lefranc, head of the school, requested that he be struck off the flying personnel list because of his incapacity: "*He is very reckless and unconscious. He does not listen to the recommendations of his chief pilot. (...) Finally, as he is very scatterbrained and reckless, and neglects the most elementary precautions, I think it would be dangerous to let him continue flying.*" Lefranc will not be followed by his hierarchy, since Robert de Bonnefoy finally obtains his military pilot's license (n°1016) on May 31, 1915.

He arrived at the general aviation reserve at Le Bourget on June 4, 1915, and left a week later for the 1st bombing group (VB 101 squadron) based on the Malzéville plateau, near Nancy, on Voisin LAS. From there he participated in several strategic bombing missions over Germany during the rest of 1915, taking part in attacks on Karlsruhe, Trier, Dillingen, Pechelbronn, and twice on Saarbrucken. During this period in the bombardment, he was promoted to the rank of maréchal des logis on August 21, 1915.

After eight months on the front, he was sent back to the rear and became an instructor at the Avord school on January 21, 1916. A position that he did not like because he wanted to fight: on April 12, he managed to join the group of training divisions (GDE) from where he was assigned to the new fighter escadrille N 68 stationed in Lorraine.

He stayed there for a short time: on May 25, 1916, he was assigned to escadrille N 65 which was fighting at Verdun and where ace Charles Nungesser was serving. It was there that he scored his first certified aerial victory by shooting down a German plane at Samogneux, less than 10 km north of Verdun, on July 2, 1916.

The N 65 then moved on July 7th 1916 to the Cachy field in the Somme, where the British troops supported by French troops tried to break through the German lines. The air battles were numerous and the German air force was both outnumbered and outgunned. Several French fighter pilots made an impressive record of achievement; he was one of them.

Promoted to the rank of adjudant on July 16, 1916, he shot down two aircraft around the town of Brie on July 27, 1916, of which only one was certified, and another one, not yet certified, at an unknown date in July. On August 3, 1916, he scored his third official success against an aircraft but was wounded the same day in the right arm by shrapnel while attacking an enemy captive balloon at low altitude.

A fall with serious consequences

Quickly healed, on August 23, 1916 he shot down an LVG C at Marchélepot at 10:30 am, his 4th official success. But the next day, August 24, he was shot down in aerial combat at 10:45

Above: Portrait taken after his first mention in dispatch, on June 1916 while flying in an Voisin LAS bomber in VB 101.

am over Combles. With the controls cut, his Nieuport fighter crashed into the British trenches. Losing consciousness for several minutes, he got up wounded in the face and chest.

Decorated with the military medal, he was evacuated to a field hospital where he remained for two weeks. Even more than the physical wounds, it was his mental state that was permanently affected. He was to suffer from psychiatric problems throughout his life, being perpetually in a cyclothymic state characterized by alternating depressive phases and manic episodes of excitement.

He was discharged from the hospital on September 8, 1916, and returned to his squadron, but he never returned to his normal state. His doctor noted that: "*When he went up in the airplane, he had minutes of anxiety, understandable, but which he knew were more intense than those of his comrades "because I feel more intensely, more painfully than they do*"".

With his 4 certified victories and two probable ones, he remains one of the best pilots of his squadron behind the great ace Nungesser. He was then entrusted with the command of one of the very first SPAD VII (n°121) delivered to the unit, on board of which he was to win new successes. On October 1, 1916, he attacked and claimed no less than three enemy aircraft: the first at 9:00 am at Puzeaux, the second at 11:10 am at Rancourt and the third at 2:10 pm on the Bois de Vaux. None of them was certified, although the German archives tell us that Leutnant Ernest Reuter, of the FA 2, died that day at Rancourt.

On November 4, 1916, he scored his 5th official success at 3:50 pm against a plane that was shot down in collaboration with Lt P. Sauvage, of escadrille N 112, on the Hospital farm. This victory made him an ace and earned him a mention in the communiqué to the armies on November 13. He became famous because the national newspapers published his biography.

However, as the aerial workbook of the N 65 squadron attests, his flights became rare. He confided to his doctors that he suffered from black spots appearing on his field of vision. Promoted to the rank of second lieutenant on a temporary basis on March 18 (confirmed on a permanent basis on April 4), he was transferred on March 26, 1917 to the N 84 squadron stationed in the Oise.

But it would be a long time before he could return to combat. In the spring, he suffered a period of depression that lasted three months, during which he cried morning and night and separated from his fiancée.

His condition improved with the summer. He was made a knight of the Legion of Honor on July 10, 1917. Then, on August 21, he scored his 6th and last official success by shooting down an enemy fighter plane – probably Leutnant Hugo Geiger, of the Jasta 34b, who was killed in the area that same day, according to German archives – on Consensoye, north of Verdun.

Out of the fight

He remained at N 84 until November 12, 1917, when he officially went on leave. At the end of this leave, on November 29, he entered the Viry-Châtillon medical center. This was followed by a long period of inactivity lasting several months, during which he married Miss Aline Millevoye, the daughter of a lawyer from Lyon.

He returned to the front on April 16, 1918 to the SPA 15 squadron, in which he made only a short passage (and a priori no flight) since he was transferred on April 21 to the BR 131 squadron, a bombing squadron on Breguet 14. This was an unusual transfer for a fighter: he left on April 25 for the GDE to learn to fly the bomber, and returned on May 7. But on May 15, he was transferred to the St-Cyr center to manage a repair shop.

He was in fact unfit for combat because of his psychiatric problems which affected his behavior. His marriage to his wife soon broke down: considering her to be of mediocre intelligence, he lived apart from her after a few months.

A career between garrisons and psychiatric hospitals

He officially returned to the front on September 4, 1918 to the SPA 23 squadron where his superior described him as "very tired" and tried to have him transferred. He remained in the army after the armistice, as an active soldier, being promoted to lieutenant on December 25, 1918. In one of his phases of overactivity, he multiplied his requests for transfer and on February 1, 1919 he was assigned to the V 557 squadron in Morocco. Four months later, on May 31, 1919, he volunteered to go to Russia and was placed at the disposal of the minister for the Siberian military mission, which he did not join.

On September 3, 1919, he was assigned to the staff of the 1st fighter squadron, then on January 1, 1920 to the 1st fighter regiment in Lorraine, becoming the acting commander of the

Above: Robert de Bonnefoy in front of a SPAD VII.

Above: Portrait taken in late 1916 for Jacques Mortane's magazine *"La guerre aérienne illustrée"*.

105th squadron. While strikes were taking place in the steel industry, the army helped maintain order and he was reported by the commander of the 162nd infantry regiment for having effectively helped repel striking attackers.

The Lorraine basin did not seem to suit Lieutenant de Bonnefoy, who again requested a transfer and was posted on October 30, 1920 to the 35th Aviation Regiment of the Levant in Syria. He returned less than two months later, on December 20, officially for health reasons. According to his doctors, it was because he did not like the work he was doing. In fact, he suffered from depression: he was evacuated for "psychasthenia" and poor general condition.

When he returned to France, he stayed in Saint-Mandrier and Tamaris-sur-Mer near Toulon, where he felt very irritable. He then rested for two months in Divonne-les-Bains, near Geneva, where he seemed to get back on track.

On April 5, 1921, he was assigned to the 2nd fighter aviation regiment in Strasbourg, where in October 1921 he took part in maneuvers and made many night flights on his Nieuport 29, which earned him praise from his superiors. But these flights caused him serious sleep disorders and he relapsed into his psychiatric problems. On January 17, 1923, he came to blows with another officer, Lt Bouvet, over an argument their respective wives had. The two officers went so far as to fight a duel and exchange fire, fortunately without damage on either side. They were both transferred for disciplinary reasons, De Bonnefoy landing on February 1, 1923 at the 1st Aviation Regiment of Thionville.

In this new garrison, he had nothing but problems, being constantly offended by the other pilots whom he considered as vulgar, and quarreling with his father, then a cavalry colonel. He was transferred again to the 2nd aviation regiment on November 25, 1923, and then, in April 1924, he suffered from what the doctors called a serious manic attack, which led to his first placement in a psychiatric hospital.

He came out of it in a better state and was transferred on August 25, 1924 to a fighter group of the 32nd aviation regiment, being promoted to the rank of captain on March 24, 1925. On June 20, 1925, he was in the 2nd aviation regiment, assistant captain to the commander of the 2nd group. He became acting commander of the 1st squadron of the group on November 1, 1925, and then took over the command of the group on February 1, 1926, becoming the deputy commander of the 2nd group on November 5, 1926. Promoted to Officer of the Legion of Honor on December 23, 1927, his marriage was better and his wife gave birth at that time to a daughter named Arlette.

But he had a relapse on October 12, 1929, during a leave in

Paris where he hit an old man in a fit of dementia and ended up at the police station. He was again placed in a psychiatric institution following this second serious manic episode.

His case was then examined by the military authorities in a reform commission which deemed it necessary to place him in a position where he would have neither physical nor intellectual effort to produce. On May 25, 1930, he was seconded to the historical service, where he was promoted to the rank of major on May 1, 1931.

However, a new placement in a care institution during the year led the army to retire him on December 7, 1931, even though he had accumulated 1,800 hours of flight time by day and 100 by night. After becoming a reserve officer, he had another stay in a psychiatric institution in 1934, which led the reform commission of Orleans to propose that he be discharged with a 100% pension for "*manic psychosis, chronic depression in the cyclothymic phase*". He was interned again in 1936 and 1938.

Facing Page: Commandant Féquant, chef of GC 13, gives a decoration to Robert Bonnefoy in St Pol-sur-Mer, on September 1917 during the battle of Flanders. He recommended him for a mention in dispatch for his 6th victory, but, probably because for his behavior problems, he never received it.

Recalled to active duty during the Second World War on September 2, 1939, he was quickly sent home on November 28. A medical commission dated May 31, 1940 noted that he was "*depressed, asthenic, remorseful, and suicidal*". He was definitively discharged from the air force on February 7, 1941, after a new medical finding of cyclothymia with manic episodes.

He did not play any role during the Second World War, which he seems to have spent on the family property in the Jura, and died on September 27, 1946, when he drowned while swimming in the Ain River.

Robert De Bonnefoy Victory List

N°	Date	Unit	Plane piloted	Plane claimed	Place	Notes	German casualties	Confirmation
1	1-Aug-16	N 65	Nieuport	Enemy plane	Samogneux			
2	2-Jul-16	N 65	Nieuport	Enemy plane	Brie			Citation OA 25-08-1916 - JORF 22-09-1916 p.8386
P1	27-Jul-16	N 65	Nieuport	Enemy plane				
P2	27-Jul-16	N 65	Nieuport	Enemy plane		Precise date unknown		
3	31-Jul-16	N 65	Nieuport	Enemy plane	Maurepas			Citation OA 25-08-1916 - JORF 22-09-1916 p.8386
4	3-Aug-16	N 65	Nieuport	LVG C	Marchélepot			Citation OA 05-09-1916 - JORF 05-12-1916 p.10512 "4th plane"
P3	23-Aug-16	N 65		Enemy plane	Puzeaux			
P4	1-Oct-16	N 65		Enemy plane	Rancourt		FA 2, Ltn Ernst Reuter killed in Rancourt	
P5	1-Oct-16	N 65		Enemy plane	Bois de Vaux			
5	1-Oct-16	N 65		Enemy plane	Ferme d'Hopital	With Lt P. Sauvage (N 112)		Citation OA 12-11-1916 - JORF 06-04-1917 p.2706 "5th enemy plane"
P6	4-Nov-16	N 65		Enemy plane				
6	24-Dec-16	N 84		Enemy plane	Consensoye		Jasta 34b, Ltn Hugo Geiger killed N. Verdun	GC 13 documents

Fernand Bonneton

First names: Fernand, Régis.
Born: 20-05-1890 in La Coucourde (Drôme).
Father's profession: Farmer.
Military pilot brevet: 15-09-1916 (n°4489).
Units: F 7 (Romanian), N 1 (Romanian), SPA 69.
Army communiqué: No.
Total score: 9 victories (5 Drachen), 1 probable.
Died: 24-06-1922 in Brussels-Evere (killed in a flying accident).

Above: War portrait of Fernand Bonneton.

Fernand Régis Bonneton was born on May 10, 1890 in the village of La Coucourde, a village located along the Rhone River just north of Montélimar, where his father was a farmer and his mother a housewife, according to the civil register.

The young man was quite tall for his time (1.75 m) and, without obtaining the school certificate, graduated from the local school with a good level of education. He did not see himself becoming a farmer like his father and decided on September 28, 1908, once he had passed his eighteenth birthday, to enlist for four years in the army by going to the recruitment office of the Montélimar town hall. He was incorporated into the 12th regiment of dragoons as a private – at the time, the cavalry regiments were the most prestigious. His unit leader was Captain Houdemon, an officer who would later become a general of the Air Force and who would always follow the young soldier's career with kindness. Fernand Bonneton served with dedication and was promoted to brigadier on April 27, 1909, then to maréchal des logis on March 24, 1911. On March 22, 1912, he extended his enlistment by one year, and then on September 24, 1913, he extended it for two more years.

On February 24, 1914, he was promoted to the position of maréchal des logis fourrier. When the war broke out, he was engaged in the front line with his regiment, which was part of the units executing the offensive in Lorraine planned by General Joffre, which ended in disaster. Fernand Bonneton received his first wound on August 29, 1914, hit in the right leg by shrapnel while observing the enemy lines. After the battle of the Marne, the war of movement ended at the end of September and the 12th Dragoon Regiment was transformed into an infantry regiment, with the horses left behind and the cavalrymen taking guard in the trenches.

He received a citation in the order of the brigade on October 8, 1914 and was promoted to *maréchal des logis-chef*

on October 18, 1914. He was then cited in the order of his regiment on November 4, 1914, for his dedication in caring for the wounded of his unit. He left the unit at the end of the year after a new injury and volunteered to serve in an infantry regiment, joining the 82nd RI in Argonne where he was promoted to second lieutenant on a temporary basis on March 22, 1915. It was also during this period that he met his fiancée, Miss Elisabeth Weiser, a nurse at the front. Fernand Bonneton was confirmed in his rank on March 1, 1916, after more than a year spent in the trenches, where he received two commendations in the divisional order, on October 6, 1915 and April 7, 1916, the text of the latter being: "*Officer of proven bravery. With the grenadiers of the battalion, rushed at the head of his men into the funnel dug by a German mine. Was wounded during the occupation of this funnel.* "

However, this latest wound (his 4th) rendered him unfit for the infantry. In order to continue fighting, he volunteered for the air force and was immediately accepted as a student

Left: Post-war portrait of Fernand Bonneton – without his moustache.

Below: Capitaine René Chambe, commanding officer of Romanian escadrille N1, inspects a Nieuport 11 in early 1917 in Moldavia. Fernand Bonneton was affected to this unit and obtained 2 victories on the Romanian front. (Passot-Seguin collection)

Above: Back to France and escadrille SPA 69 in 1918, Fernand Bonneton poses in front of a SPAD VII.

pilot on April 22, 1916. He then went to the Juvisy school and obtained his military pilot's license (n°4489) on September 15, 1916 on a Caudron plane, then left to perfect his skills at the Châteauroux school.

At the end of his training, unlike many pilots, he did not join a squadron at the front, but was assigned on October 19, 1916 to the Romanian mission, a group of pilots who were to form the backbone of the new Romanian air force being armed by France. In November 1916, he was in Salonika with a group of French pilots preparing to join the Romanian air force on Farman aircraft. The raid was finally cancelled and the pilots joined their assignment via Russia, which they reached at the port of Murmansk.

Promoted to lieutenant on March 22, 1917, Fernand Bonneton was assigned to the Romanian F7 squadron equipped with Farman F40 aircraft plus some Caudron G 4, and stationed in Moldavia. He flew many missions on the Carpathian front where the Romanian army was taking position. On August 8, 1917, flying a Caudron G 4 he did not hesitate to attack an enemy fighter that was going to attack a balloon, and allowed his gunner to shoot it down. The victory was recognized by a citation in the army order published in the official journal on December 24, 1917. This success allowed him to be assigned to a local fighter unit and he joined the N1 squadron of the Romanian Air Force. On October 30, 1917, he won another victory aboard a Nieuport by shooting down an enemy plane that was threatening a Romanian reconnaissance plane.

After the armistice requested by Russia, which in fact precipitated the collapse of the Romanian front, Bonneton had to return to France by fighting his way through the tumult of the Russian revolution. He returned in April 1918 and, after a period of rest, was assigned on June 26, 1918 to the SPA 69 squadron stationed in Oise.

It is there that he would quickly increase his hunting record at the controls of his SPAD decorated with a cat's head, the unit's insignia. He discovered a specialty in attacking captive balloons: after having shot down a plane between Noyon and Ribécourt on July 10, 1918, he set fire to a first Drachen on August 1, 1918, which earned him the Knight's Cross of the Legion of Honor on August 3. He destroyed a second Drachen on the 15th of the month. His third destroyed balloon on August 22 gave him cold sweats as he passed with his SPAD through the balloon's fire cloud and returned to his base with debris from the envelope of his prey stuck along his wings. September 22 was a great day because he won a "double", setting fire to a Drachen then shooting down a Fokker D VII in the same day. Rested, he only aspired to return to the front to obtain a total of 10 victories and have the honor of seeing his name appear in the communiqué to the armies. He failed by a small margin: on October 22, 1918, he claimed two Drachen in the Laon region, but only one was approved.

He ended the war with 9 certified victories obtained in 320 flying hours. Maintained in the SPA 69 squadron, of which he took command in the second quarter of 1919, Fernand Bonneton was promoted to the rank of captain on April 4, 1919. He then left for the military mission in Poland, where he was promoted to Officer of the Legion of Honor upon his

Above: Fernand Bonneton poses with the remnants of the Drachen he shot down on August 22, 1918 : he flew across the exploding balloon with his SPAD!

return in June 1920, after a new injury.

He was then assigned to the 1st Fighter Aviation Regiment in Thionville under the command of Colonel Houdemon, where he commanded the 3rd group, and married his fiancée Elisabeth Weser on December 20, 1920, who gave birth to a daughter. Piloting the new Nieuport 29s, he led a delegation of five French pilots to the Brussels-Evere international air show. It was there that he lost his life on the afternoon of Saturday June 24, 1922, during a balloon shooting exercise, when his plane crashed under the eyes of his wife.

SPAD XIII (n°83??) of Lieutenant Fernand Bonneton, SPA 69, 1918.

Above: Fernand Bonneton posing in front of his SPAD XIII. Note the cat's head, insignia of escadrille SPA 69.

Left: Capitaine Fernand Bonneton, then commanding officer of escadrille SPA 69, at the command of a Nieuport 29 fighter after the armistice.

Fernand Bonneton Victory List								
N°	Date	Unit	Plane piloted	Plane claimed	Place	Notes	German Casualties	Homologation
1	8/8/17	F 7*	Caudron G4 n°1814	Fighter	N. Marasheshti (Romania)	With S/Lt Naderval as observer.	Halberstadt D.II of FA 42 lost on the sector	Citation OA 10-08-1917 - JORF 24-12-1917 p.10530
2	10/30/17	N 1*	Nieuport	Two-Seater	Magura / Tirgul Ocna (Romania)	With Cpt Charles Micheletti		Citation OA - Romanian communiqué
3	7/10/18	SPA 69	SPAD	Enemy plane	Noyon-Ribécourt			Citation OA(LH) - JORF 05-08-1918 "3rd victory"
4	8/1/18	SPA 69	SPAD	Drachen	Aisne			Citation OA - JORF 05-07-1919 p.6882 "7th victory"
5	8/15/18	SPA 69	SPAD	Drachen	Pontoise			
6	8/22/18	SPA 69	SPAD	Drachen	Vauxillon			Citation OA - JORF 26-05-1919 p.5423
7	9/2/18	SPA 69	SPAD	Drachen	Pontavert			Citation OA - JORF 05-07-1919 p.6882 "8th victory"
8	9/2/18	SPA 69	SPAD	Fokker D VII	Beaurieux		Ltn Gottfried Clausnitzer (1v), Jasta 72, who shot down a french ballon from 72e Cie.	
9	10/22/18	SPA 69	SPAD	Drachen	Laon			Citation OA - JORF 15-04-1919 p.3962 "9 victories"
P1	10/22/18	SPA 69	SPAD	Drachen	Laon			

* Romanian

Charles Borzecki

Name: du Bois d'Aische (Count).
First names: Alexandre, Charles, Bronislas.
Born: 04-11-1881 in Paris (10).
Father's profession: Merchant.
Military pilot brevet: 10-03-1918 (n°12082).
Units: C 43, N 62, N 79.
Army communiqué: No.
Total score: 5 victories.
Died: 06-03-1957 in La Flèche (Sarthe).

Above: Portrait of Charles Borzecki, taken between December 1915 and march 1916 according to the number of palmes shown on his Croix de guerre.

A young engineer specialized in aerial photography

Alexandre, Charles (his usual first name), Bronislas Borzecki was born on November 4, 1881 in Paris where his father, Bronislas (aged 31) was a merchant. His mother, born Caroline Lorch, was 39 years old and the couple lived at 46 boulevard de Magenta, in the 10th district of Paris. He came from a rather wealthy family, which employed a servant who was a witness on the child's birth certificate, along with the building's janitor.

The young brown-haired, blue-eyed boy grew up to be a giant of 1m80, an exceptional height for the time, and studied technology to become an electrical engineer. On October 15, 1901, the year he turned 20, he had to fulfill his military obligations: he was drafted into the 15th foot artillery battalion as a private. The unit was based in Condé sur l'Escaut, near Lille on the Belgian border. Borzecki quickly gained rank, being appointed to the rank of brigadier on March 1, 1902, and because of his education became chief of the photo-electric posts on March 15, doing an internship at the photographic school in Le Havre.

After returning to civilian life on November 1, 1903, he continued to practice his specialty in the reserve and voluntarily completed a one-month internship on October 3, 1904 at the photographic section of Chalais-Meudon, where he learned to take aerial photographs in balloons. On August 25, 1905, he was attached to the 6th foot artillery regiment, a fortress unit whose batteries were scattered along the border in Lorraine.

Married in Paris on March 26, 1906 to Miss Marguerite Lombard (the couple did not have children), he worked as an electrical engineer in civilian life when the general mobilization called him back into the army on August 2, 1914, where he was assigned to the 3rd foot artillery regiment stationed in Brest, which was to be engaged in Belgium during the first fighting, in Maubeuge and then Ypres. Charles Borzecki was appointed *maréchal des logis* on September 15, 1914, and because of his qualifications in aerial photography, he was appointed on November 1, 1914 as a photographic section operator in the aviation of the 6th French army stationed in the Oise region.

Facing Page: Portrait of Charles Borzecki, taken between December 1915 and march 1916 according to the number of palmes shown on his Croix de guerre.

Left: Lt. Charles Borzecki makes the fool on top of the fuselage of this Nieuport 14. Around the plane, 3 two-seater pilots of escadrille N 62: Sgt Edouard Pinot, Sgt John Huffer (American pilot) and Sgt Gabriel Hébert. Borzecki scored 3 victories with the last two pilots. The photo is dated December 18, 1916. (Albin Denis collection)

Becomes an observer in the air force

As such, he made several flights as an observer on the various aircraft of the different squadrons of the 6th Army (MS 3, MF 62, MF 16, C 10, C 43, C 46, MF 19, MF 33), being frequently targeted by enemy flak, which caused him two head wounds from shrapnel - the first one at Coucy-le-Château on March 10, 1915, and the second one on April 2, 1916 at Clermont in the Oise. Rewarded with a citation in the army order on September 26, 1915 for a series of observations conducted from September 9 to 12, 1915, it was not until June 28, 1916 that he was officially attached to a squadron, the C 43, being promoted to the rank of aspirant (officer cadet) two days later and taking the position of photo section leader. It should be noted, however, that Charles Borzecki had already flown several photographic reconnaissance missions on board C.43 aircraft, which were decorated with a dragonfly. His Caudron

Above: Pilots and observers of escadrille N 62 pose in front of a Sopwith 1A2 in Fismes airfield on March, 1917. On the ground, left to right: unknown, Lt Charles Borzecki, a RNAS officer, Cpt François Coli (commanding officer, disappeared during the Atlantic crossing attempt with Charles Nungesser in 1927), Adj Paul Tarascon (12 victories), S/Lt Paul Favre de Thierrens (6 victories), S/Lt Robert de Francq (pilot), Lt Pierre l'Huillier (pilot). Behing, standing on plane's wheel, is Lt Paul Dupont (pilot). (Albin Denis collection).

G 4 was always escorted by two Nieuport 10 fighters.

Borzecki's official affection coincided with the beginning of the Battle of the Somme in which the 6th Army took part. He distinguished himself from the very first day of the battle, as on July 1, 1916, he had to defend himself with machine guns to repel an enemy fighter attack, which he did again two days later. On July 25, 1916, aboard a Caudron G 4 whose pilot was unknown, he shot down an enemy two-seater in the Combles area – German records indicate that day the loss of a Kasta 11 crew, Uffz. Georg Roder and Lt. Friedrich Zilling, killed at Bapaume near Combles. He was awarded the Military Medal on August 4th – not for the aerial victory, but for the previous fighting.

He continued to distinguish himself on the Somme front, being transferred at an undetermined date to the N 62 squadron, an army squadron whose main mission was hunting but which had two-seater planes intended for major reconnaissance missions. He succeeded in a particularly long one on September 17, 1916, which took him 100 km into the enemy lines and earned him a third citation in the army order on September 28. These missions were not without danger, as the enemy was on the prowl, but he proved to be just as good at defending himself with a machine gun as he was at taking photos. On October 10, 1916, on a Sopwith 1A2 flown by Sergeant John Huffer (an American volunteer), he scored an aerial victory when attacked by four enemy aircraft and managed to shoot down one of the attackers east of Peronne, putting the others to flight. The fighters of the N 62 squadron escorting the Sopwith were dispersed during the fight, and Huffer and Borzecki decided to cross the enemy lines alone and accomplished their mission. The same crew did it again on November 23, 1916, when they faced eight enemy aircraft and shot down one near Ypres, while one of the escorting Nieuports shot down a second.

Appointed to the rank of second lieutenant on December 14, 1916, and made a knight of the Legion of Honor a week later as a reward for his last two aerial victories, Charles Borzecki continued his missions in 1917 and flew with MdL Gabriel Hébert as a pilot on February 10, 1917, where he flew a long-distance photo reconnaissance flight north of Laon and was attacked by three German aircraft. One of them was shot down in the vicinity of Etouvelles, south of Laon, and the two others broke off the fight. The Sopwith 1A2 n° 22 of the N 62 was hit by 5 bullets, one of which cut the elevator control. The

Left: Charles Borzecki in front of a SPAD of SPA 79, his last war unit. The wolf's head is the unit insignia.

victorious crew did not return home until their mission was accomplished.

On February 25, 1917, Charles Borzecki won his 5th and last air victory in a two-seater flown by Corporal Frédéric Fournier, a young pilot who was performing one of his first reconnaissance flights. Attacked by three enemy fighters, Borzecki shot down one of them which fell south of Pinon.

He received two citations in the army order on March 8 and 19, 1917 for his two air victories, the last one stating that he had achieved a total of 5 air victories. Curiously, he was not mentioned in the Army's communiqué, as is traditional for any pilot who has obtained five air victories, but this was also the case for the machine gunners of the C 46 squadron. He was therefore not famous, except for a few articles in Jacques Mortane's specialized magazine "*La Guerre Aérienne Illustrée*" (The Illustrated Air War), which pointed out this injustice and awarded him the title of forgotten ace.

However, the war continued and he continued his missions with the N 62, flying a two-seater SPAD XI at the end of 1917. But after all these missions as an observer, he wanted to take the controls himself and learn to fly in a squadron in October 1917, then regularize his situation by going to the Groupement des Divisions d'Entrainement in December (he was confirmed in his rank of second lieutenant on the 10th of the month), then obtaining his military pilot's license (n°12082) on March 10, 1918 – at the age of 37, which makes him one of the oldest pilots to be licensed during the First World War.

Having received a last citation (his 6th) in the army order on January 6, 1918 for the 500 or so photographs of the lines taken at the N 62 squadron, he went back to combat at the N 79 squadron, another army squadron, on March 23, 1918, until May 8, 1918, when he was appointed instructor at the CIACB as head of the photographic section. Appointed to the rank of lieutenant on a temporary basis on June 9, 1918, he finished the war with 768 hours of combat flying time, plus another 144 hours of training.

In the Indochina air force

Demobilized on March 9, 1919, he retired to 56 rue de Lancry in Paris and worked as an engineer for the L. Aubert film company, where he was in charge of the equipment department. A valuable technician, he was the author of a patent registered on November 15, 1919 for a device allowing the use of a single lantern and a single light source for a two-projector station. This device did not require any modification of the equipment, nor of the protective casings, eliminates the cumbersome tables, and can be used in cabins with very small dimensions.

He did not make a career in this company, from which he resigned at the end of the year to embark in December 1919 for Indochina, at the invitation of the government of the colony which called upon his services as a contractual employee to carry out a topographic survey of the country.

As direct assistant to Major Glaize, head of the local military aviation, he organized the military photographic service in Indochina, even though he was under civilian status.

In parallel with this function, he contributed to the creation of the entire air infrastructure of the colony and, as such, carried out numerous flights on Breguet 14s or other aircraft, which caused him to have several accidents while landing on rudimentary runways, including a plane crash on June 5, 1921 on the Bach Maï field. In 1925, he distinguished himself by carrying out a complete photographic survey of the Mekong.

Numerous testimonies converge to underline the dynamism and the quality of the work of Charles Borzecki, who registers in October 1922 a new patent for a device ensuring the automatic and continuous retraction of the plates in the aerial cameras. He hunted wild animals with the pilot and 14–18 ace Pelletier Doisy during his Paris-Tokyo raid in 1924, then personally accompanied General Hirschauer, Inspector General of Air Navigation in 1936, on his mission to China.

Promoted to Officer of the Legion of Honor in 1923, then Commander on June 16, 1938, he was disbarred from the Air Force on April 5, 1939 with the rank of reserve captain, at the time when he was in charge of Hanoi airport.

He remained in Indochina during the Second World War, while the country was invaded by Japanese troops who did not notice this discreet retiree. At the return of the French troops at the end of 1945, he made available to the French troops all his personal archives on the aerial infrastructures, which proved to be very useful when all the installations of the colony were ransacked by the Japanese. When he left Indochina on August 9, 1946, at the age of 65, Admiral Thierry d'Argenlieu, the governor, wished to send him an official token of appreciation for services rendered. He flew a total of 3280 hours during his entire career.

Above: Portrait taken after the armistice, with all his decorations and légion d'honneur.

Right: One of the many aerial photos taken by Charles Borzecki in Indochina during the 1920s and the 1930s – here, Bach-Maï airbase.

Charles Borzecki Victory List								
N°	Date	Unit	Plane piloted	Plane claimed	Place	Notes	German Casualties	Homologation
1	7/25/16	C 43	Caudron G4	Enemy plane	Combles	Pilot: Unknown	Kasta 11 two-seater lost on the sector.	
2	10/10/16	N 62	Sopwith 1A2	Enemy plane	E. Péronne	Pilot: Sgt John Huffer		Citation OA 23-12-1916 - JORF 19-06-1917 p.4724
3	11/23/16	N 62	Sopwith 1A2 n°6	Enemy plane	Ypres	Pilot: Sgt John Huffer		Citation OA 23-12-1916 - JORF 19-06-1917 p.4724
4	2/10/17	N 62	Sopwith 1A2 n°22	Enemy plane	Laon	Pilot: Sgt René Hébert		Citation OA (LH) 08-03-1917 - JORF 07-04-1917 p.2727 "4 planes shot down"
5	2/25/17	N 62	Sopwith 1A2	Enemy plane	S. Pinon	Pilot: Brig Fournier		Citation OA 19-03-1917 - "5 planes shot down"

After returning to France with his wife, he became a member of the Association des vieilles tiges (pioneer pilots) and les gueules cassées (crippled veterans). Promoted to the rank of Grand Officer of the Legion of Honor on March 6, 1957, he died in a retirement home of this institution in La Flèche (Sarthe) on March 30, 1959 and has since been buried in the St Thomas cemetery of this town.

Caudron G4 (serial unknown) of escadrille C 43 flown by Borzecki as observer in late 1915 and 1916. Note the first symbol of the unit, a butterfly.

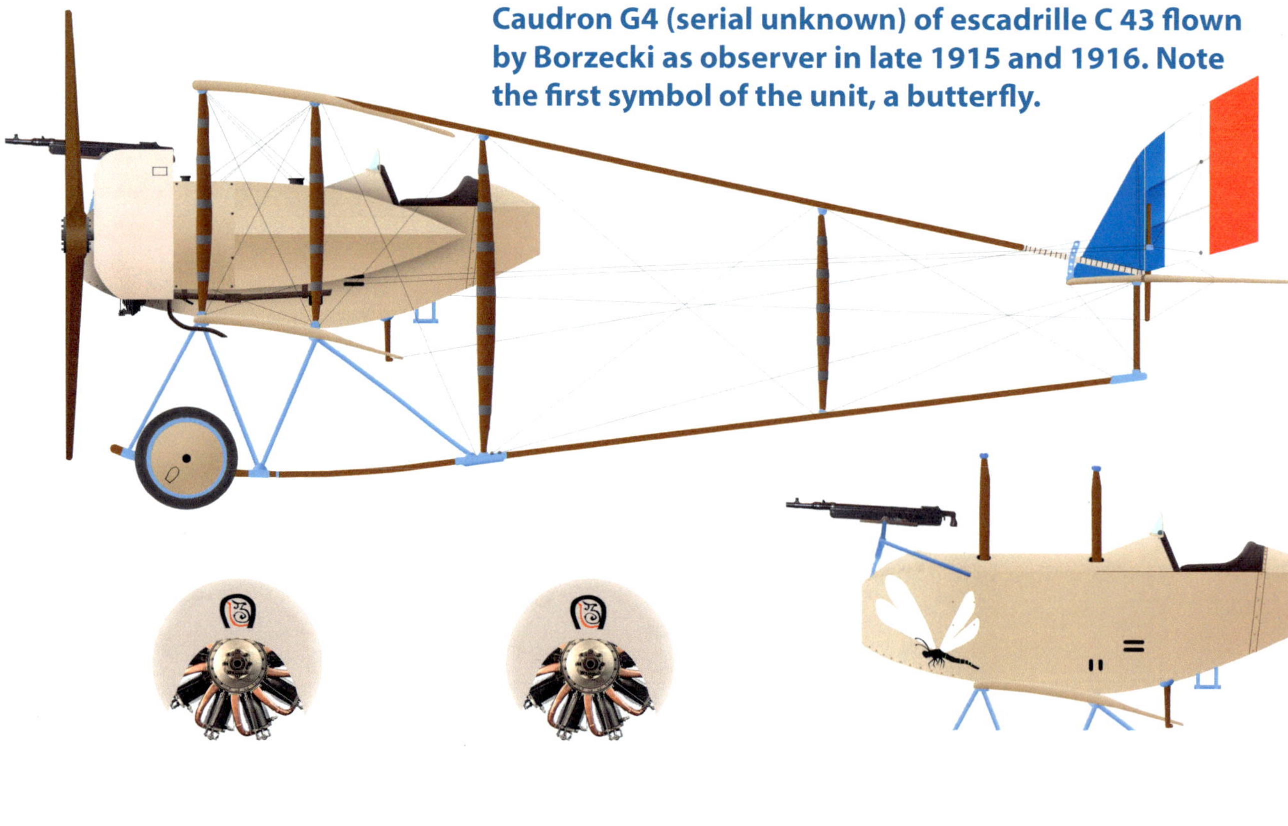

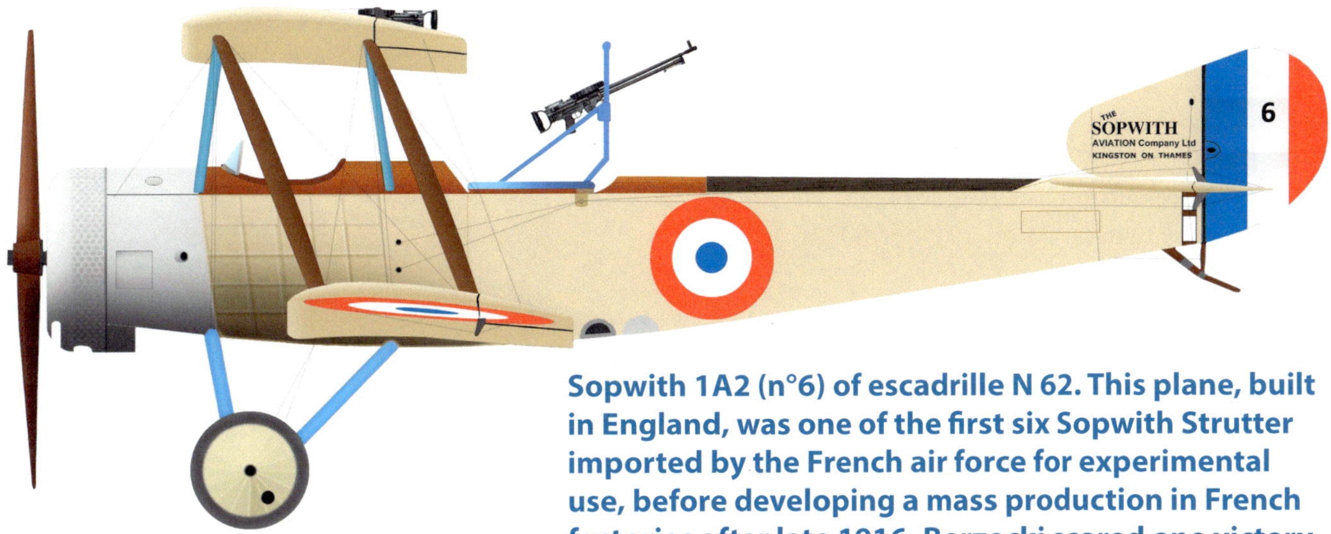

Sopwith 1A2 (n°6) of escadrille N 62. This plane, built in England, was one of the first six Sopwith Strutter imported by the French air force for experimental use, before developing a mass production in French factories after late 1916. Borzecki scored one victory as observer-gunner on this plane on November 23, 1916, with the sergent John Huffer as pilot.

One of the French-built Sopwith received by escadrille N 62. It was probably in such a plane where Borzecki scored his last victory on February 25, 1917.

SPAD XI n°53643 of escadrille N 62, Saconin-Breuil, fall 1917. Borzecki made his last missions while in N 62 on this plane.

André Bosson

First names: André, Louis.
Born: 01-06-1894 in Esmans (Seine-et-Marne).
Father's profession: Domestic servant.
Military pilot brevet: 25-09-1917 (n°8912).
Units: SPA 62.
Army communiqué: No.
Total score: 7 victories, 3 probables..
Died: 20-07-1918 in Oulchy-le-Château (Aisne).

Above: Portrait of André Bosson. The photo was altered as André Bosson received the légion d'honneur posthumously in 1922. According to French archives, he received only the médaille militaire and the croix de guerre with 4 palms (1 posthumously) and 1 bronze star.

André Louis Bosson was born on June 1, 1894 in Esmans, a town in the Seine et Marne department. His family seems to be very modest, since his father, aged 24, worked as a domestic servant and his mother, aged 19, had no profession. According to the historian Daniel Porret, his parents were Swiss citizens and lived in the town of Montereau, but André Bosson was indeed a French citizen.

When the First World War broke out, he had not yet done his military service and decided to voluntarily enlist for the duration of the war on September 7, 1914. As proof of his French nationality, he was not incorporated in the Foreign Legion, but in a unit of the regular army, the 2nd Engineer Regiment, which he joined on September 10 and where he served for almost three years, being appointed corporal on November 8, 1914, and obtaining a citation in the order of his regiment in May 1917.

Shortly thereafter, following his request, he was accepted into the air force as a student pilot on June 15, 1917. He then followed his apprenticeship by attending the Dijon schools for theoretical training, then the Juvisy flying school where he obtained his military pilot's license n°8912 on September 25, 1917, on a Caudron G3. After passing through the Avord training school, he was sent to the Pau acrobatics school and to the Groupement des Divisions d'Entrainement on December 3, 1917, while waiting for an assignment to an operational unit.

This came on December 22, 1917, when he received the order to go to escadrille SPA 62. It was there that he quickly revealed himself as a first-rate fighter and marksman. On March 9, 1918, he shot down his first enemy plane that fell near Clary-Thierret and was confirmed, plus two others that were not confirmed. Three days later, on March 12, he shot down a Fokker triplane at Crepy-en-Laonnais.

Promoted to sergeant on March 27, 1918, he again distinguished himself by shooting down three aircraft in quick succession on May 27, 28, and 30, 1918, followed by two others on June 2 and 3, 1918. He was awarded the Military Medal on June 16 and the Legion of Honor shortly thereafter by a military administration that had difficulty keeping up with his successes.

Unfortunately, he fell in aerial combat on July 20, 1918, on his SPAD VII n°5325 nicknamed "Nénette" near Hartennes, south of Soissons. Hit by several bullets in the stomach and legs, he had the strength to land his SPAD VII in the German lines. Three German soldiers quickly arrived on the scene and tried to treat André Bosson, who died shortly afterwards.

Right: Corporal André Bosson, from 2nd engineer regiment, posing in 1915 among the ruins of the church of Craonne.

Below: 5th from the right, sergent André Bosson among SPA 62 pilots. 4th from left is ace Charles Quette (10 victories).

André Bosson Victory List							
N°	Date	Unit	Plane piloted	Plane claimed	Place	Notes	Homolo-gation
1	3/9/18	SPA 62	SPAD VII n°5325	Enemy plane	Clary-Thirret		Citation OA 15-03-1918 - JORF 10-04-1918 p.3080
P1	3/9/18	SPA 62	SPAD VII n°5325	Enemy plane	Clary-Thirret		
P2	3/9/18	SPA 62	SPAD VII n°5325	Enemy plane	Clary-Thirret		
2	3/12/18	SPA 62	SPAD VII n°5325	Fokker Dr.I	Crepy-en-Laonnais		
3	5/27/18	SPA 62	SPAD VII	Two-seater			Citation OA (MM) 16-6-1918 "3th, 4th, and 5th enemy plane"
4	5/28/18	SPA 62	SPAD VII	Two-seater	Fismes	With Adj Léon Blanc, Adj Léon Veau, and Sgt Abel Dupont	Citation OA (MM) 16-6-1918 "3th, 4th, and 5th enemy plane"
5	5/30/18	SPA 62	SPAD VII	Albatros D	Ivord		Citation OA (MM) 16-6-1918 "3th, 4th, and 5th enemy plane"
6	6/2/18	SPA 62	SPAD VII	Albatros D	Fère-en-Tardenois		Citation OA 12-07-1918 - JORF 09-12-1918 p.10598 "6th and 7th airplane shot down"
7	6/3/18	SPA 62	SPAD VII	Albatros D.V	Soissons / Aisne river	With Adj Favre de Thierrens	Citation OA 12-07-1918 - JORF 09-12-1918 p.10598 "6th and 7th airplane shot down"
P3	7/19/18	SPA 62	SPAD VII	Enemy plane	Ourcq valley		
Casualty column omitted due to lack of data.							

Above: Sergent André Bosson in front of his SPAD VII nicknamed "*Nénette*" ("girl", or "chick").

SPAD VII n°5325 *Nenette* of Sgt André Bosson, SPA 62, summer 1918.

Léon Bourjade

First names: Jean, Pierre, Marie, Léon.
Born: 25-05-1889 in Montauban (Tarn-et-Garonne).
Father's profession: Civil servant – tax collector.
Military pilot brevet: 17-06-1917 (n°7070).
Units: N 152.
Army communiqué: 23-07-1918 for 10 victories.
Total score: 27 victories (26 Drachens), 2 probables.
Died: 22-10-1924 in Yule Island (Papua New Guinea)

Above: Portrait of Léon Bourjade as a novice priest, probably taken in Fribourg (Switzerland) in 1912 or 1913. (Coll Albin Denis)

A young man won over by religious faith

Jean Pierre Marie Léon Bourjade, who would retain Léon as his usual first name, was born on May 25, 1889 in Montauban, the prefecture of Tarn et Garonne. He came from a practicing Catholic family of which he was the sixth child. His father, Mr. Maurice Bourjade, was like his grandfather a tax inspector, but came from a family that had given several soldiers to the royal French armies. Having a relatively comfortable income considering France at the end of the 19th century, the Bourjades would settle in a house near Montauban, in Honor-de-Cos, in a hillside near the Aveyron.

The young boy, blond, raised in the Catholic, even monarchist tradition, proved to be gifted in all sorts of sports and was able to do a somersault with his feet. But he turned out to be a very shy and rather dreamy child. According to a family account, he was already playing the priest by celebrating masses with his three young sisters as audience.

As a teenager, he was sent to the Catholic college Saint-Théodard in Montauban, at a time when French society was in a state of intense debate about religion. The Catholic Church was losing its audience in large parts of society and was threatened by the growing influence of the Freethinkers' society, whose ideas were those of politicians from the left to the center-right, and more particularly the Radical Party which came to power in 1902 (bloc des gauches). The policies of the governments that came out of this bloc expelled the religious congregations from France in 1903 and then suppressed the teaching congregations in 1904 before the law of separation of church and state was adopted in 1905 under the initiative of Aristide Briand and the head of the government Emile Combes. During the inventories of church property, fights involving the police force took place in the most religious regions of France, as was the case in Montauban. These events had an indirect consequence on his schooling, since in 1904 his school took in pupils from the dissolved Catholic schools. One of them, from the teaching congregation of the Sacred Heart Fathers of Issoudun, convinced him and his older brother Victor to become missionaries. Like all teenagers of their age, the two Bourjade brothers read adventure stories by Jules Verne and Fenimore Cooper, and were fascinated by their comrade's

Right: Second Lieutenant Léon Bourjade photographed in April 1917. He is at this time a student pilot.

Below: PFamily photo of the N 152 squadron at its beginning in the fall of 1917. From left to right, front row: Sgt Jean Hahn and Cal Marcel Denis. Second row: Sgt Ernest Maunoury, S/Lt Charles Lafargue, Lt Charles Lefèvre (squadron leader), S/Lt Léon Bourjade and Cal Philippe Gresset. Third row: MdL Sylvain Dufour, MdL-Chef Alexis Lendresse, MdL Gontran de La Marque, Adj Albert Poulain, Adj Christian Schlumberger, Brig Louis Giband and Cal René Vandendorpe.

Above: Bourjade among his family, showing the Sacré Coeur banner he installed in his plane.

account of the religious missions to the other side of the world organized by his institution, and especially to Oceania among certain cannibal tribes. The young Léon Bourjade, described by a witness at the college as "distant and comical, an inhabitant of another planet", found his way and decided to enter the orders to become a missionary.

Novice priest

He abandoned his studies for this reason. After failing his first baccalaureate in 1907, he took it again in 1908 and passed the written exams. However, he decided not to take the oral exams and, with the consent of his parents, decided to join the novitiate of the Fathers of Notre-Dame du Sacré College in Issoudun, which had taken refuge abroad in Spain. Before his departure, he nevertheless made a pilgrimage to Lourdes, which he accomplished himself on his bicycle... He then began his novitiate at the religious institute of Canet-de-Mar in Spain, joining his older brother Victor who was already there.

It was there that he discovered in the library of the institution a book that was to be a revelation for him, "*Histoire d'une âme*" (History of a soul), the autobiographical account of a young Carmelite nun who died of tuberculosis in 1897 at the age of 24, named Marie-Françoise Thérèse Martin, better known by her religious name Sainte Thérèse de Lisieux. Threatened by the decline in religious practice at the beginning of the century, the church attempted a charismatic revival by widely publishing this little book at the beginning of the century, which was also a real publishing success (nearly one million copies distributed in 1915). The story of the young nun's ascent to faith was to arouse vocations and her tomb became a place of pilgrimage: several cases of miraculous cures were reported, including that of a certain Edith Gassion, the future Edith Piaf, who was cured of her keratitis in 1920 and devoted her entire life to the cult of the young saint. The

Above: View of the Nieuport 24 or 27 of the N 152 squadron at the Chapelle-sur-Chaux airfield, near Belfort, at the end of 1917. The Nieuport coded 2 on the right of the picture could logically be that of Léon Bourjade.

Above: Pilots of the N 152 squadron at Chaux airfield in April 1918, proudly posing behind their crocodile mascots. Bourjade, bearded, is in the center of the picture. At that time, he started his record of achievements by having already set fire to two Drachen officially confirmed to him. (Coll Albin Denis)

young 17 year-old novice priest Léon Bourjade identified with the religious journey of the Saint and devoted his whole life to an unprecedented devotion to her, addressing her personally in the diary he kept during the war, feeling that he was under her personal protection in the face of danger.

He left Spain for Switzerland in August 1909, where his religious institution moved to Fribourg in the French-speaking part of Switzerland. He continued his schooling there for almost a year, before being called back to France for his military obligations, because, being of the class of 1909, he obtained a deferment of incorporation for one year. The young man of 1m64 m was incorporated on October 8, 1910 in the 23rd Field Artillery Regiment in Toulouse. His level of education allowed him to be appointed brigadier on October 15, 1911, but not much more: his superiors were dubious, to say the least, about his whimsical character and noted that he could "in a pinch, make a piece leader". His two years of service allowed him to return frequently to the family home in Cos during his leave.

Released from his military obligations on October 1, 1912, he returned to Fribourg to continue his training as a priest where he met and befriended a student who had already been

Above: Léon Bourjade's favorite target: a German captive balloon, known as "Drachen", which were vulnerable because they were inflated with hydrogen, but very strongly defended by the enemy's flak and hunting. He officially set fire to 26 of them (winning another homologated victory against a Pfaltz D.XII), the record among French aces. (Coll. Greg Van Wyngarden)

ordained a priest named Henri Garin. On July 28, 1914, he returned to the family home in Cos to attend the wedding of his older brother, a colonial administrator in Indochina. The ceremony was scheduled for August 4... But the general mobilization came two days before and Léon Bourjade took the train to Toulouse to join his 23rd field artillery regiment.

Fighting in the trenches

From Toulouse, he immediately left for the front and on August 22, 1914, he took part in the battle of Charleroi, where the German army tried to outflank the left wing of the French army through Belgium, reinforced with English troops. The French army retreated... Bourjade then took part in the Battle of the Marne with his regiment, from September 4 to 8, 1914, where the German troops were stopped and forced to retreat thanks to the good use of artillery and aviation by the French. Léon Bourjade, at the end of the battle, was promoted to the rank of maréchal des logis and head of a 75 mm gun battery on September 24, 1914.

The war quickly got bogged down in the trenches and he had the pain of losing one of his older brothers, Paul Bourjade, captain in the 37th infantry regiment, who fell in Flanders at Bixschoote on November 16, 1914. The classic artillery was now supported by a new type of gun, the "crapouillots", 58 mm caliber trench mortars that threw their shells in oblique fire to fall vertically on the enemy trenches. Bourjade was assigned to such a section in February 1915 and experienced the hell of the front-line trenches. On several occasions, he came under fire from enemy trench mortars, or even from the imprecise fire of the French heavy artillery, which buried or pulverized his pieces and servants. He escaped each time and attributed his good fortune to the divine intervention of Saint Therese of Lisieux, as he noted in his diary.

On June 1, 1915, he was transferred to the 57th artillery regiment (to the 105th battery of 58) and participated in the Battle of Artois on September 25, 1915. He received his first commendation on October 15, 1915, in the order of the 17th army corps, at the end of this battle: "*Maréchal des logis, section commander in the front line trenches since February 1915, has constantly given his men the best example of courage, energy, and coolness. Distinguished himself in May, June, and July 1915. Charged with a very dangerous post from September 21 to 26, 1915, fulfilled the mission entrusted to him although his pieces had been repeatedly buried by enemy shells and his servants, except two, killed or wounded.* "

He then took part with his regiment in the battle of Verdun in May and June 1916, then on the 21st of that month left the front temporarily to attend the preparatory trench artillery course given at Bengy-sur-Craon, in the Cher department. He returned on July 22, 1916 and was promoted six days later to the rank of second lieutenant on a temporary basis, thus

Above: View of Léon Bourjade in front of his personal SPAD XIII serial 2262. The portrait of Saint Theresa on the front of the fuselage and the flag of the Sacred Heart fixed behind the cockpit are clearly visible.

becoming an officer. On August 2, 1916, he was transferred to the 59th artillery regiment, assigned to the 121st battery of the 58th artillery, where he was chief of the half-battery. He was then immediately assigned to the Somme where he participated in the battle until October 1916. On December 28 and 29, 1916, he came under German artillery fire and a 150 mm shell exploded nearby, burying him alive. He came out of it without the slightest injury and noted in his diary that "*the German artillery had, on several occasions, just missed me*". He even took part in a night expedition into the enemy trenches, from which he narrowly escaped.

Passage into the air force

It was at the beginning of 1917 that he had the idea of joining the air force, if we are to believe a letter addressed to his brother Victor: "*I am wasting my time in this profession which consists of killing my brother across the trenches so as not to be killed by him. Whereas in Oceania, once a missionary, knowing how to fly an airplane would be a precious asset to me.* "

He then applied to join the air force, which was accepted on March 28, 1917. From then on, he followed the curriculum of the flying schools, following the theory in Dijon, then the actual flying in Avord where he received his military pilot's license on a Caudron G3 (n°7070) on June 17, 1917 after 35h35 of flight. As he confided to the journalist Jacques Mortane, he experienced the greatest fear of his life when he narrowly escaped a collision. His diary, addressed to Sainte-Thérèse, is more precise on July 16, 1917: "*Tonight, on my first flight alone, on a Nieuport 23 meters, I was descending with the engine off, a little worried about my field, when, eight or ten meters away from me, I saw a double command coming at me. There was nothing to do but wait and hope for a providential crossing. And we did indeed cross paths. The tail of my aircraft passed under the wing of the other. But how my heart was beating! My thoughts were full of you. Yes, it was you who inspired the pilot of the other aircraft – a soul who also loves you very much – to pitch up quickly, which allowed the crossing to take place.*"

He then left for the Pau acrobatics school on July 27 to be instructed in the piloting of fighter aircraft, and then moved to the Groupement des Divisions d'Entrainement at Le Bourget on August 10 to await his assignment to an operational unit. He was then quickly assigned to escadrille N 152, based on the Vosges front near Belfort, which he joined on September 13, 1917.

Difficult integration into his unit

Created barely a month earlier, escadrille N 152 was led by Lieutenant Charles Lefèvre and set up on the Corcieux airfield on August 21, 1917. His men chose a crocodile as their emblem, which was painted on the side of the unit's Nieuport 24, 24bis, and 27 aircraft, and distinguished themselves on October 20, 1917 by forcing Zeppelin L 49 to land at Bourbonne-les-Bains. The airship, which had been disoriented by French radio jamming the previous night while returning from England, found itself in Lorraine in the morning in French lines where the pilots, warned by telephone, went to intercept it.

Bourjade did not take part in the action because he had not yet received an aircraft, and he was bored with his flute brought back from the trenches, as he wrote to his father at the end of the year: "*... for me, I still do not fly, my aircraft has not arrived and the weather is rarely good. In the opinion of my comrades, it will be a good thing if each month we fly an average of ten hours. The rest of the time*

Above: View of Charles Maheu, Bourjade's personal mechanic, riding the SPAD XIII of his boss.

there is nothing to do. Also for my flute, the period of activity has already resumed. There is a small orchestra in the squadron: piano, violins, flutes, etc... and I will try to make a number of them. I also have the beautiful intention of cultivating a little philosophy and theology. This will be just as pleasant as the cultivation of mud in the gut... Conclude that I am not to be pitied."

Seminarian among a band of twenty-year-old pilots, the integration of the young second lieutenant is not without difficulties among men who are the image of French society and include many anti-clericals. This fact is mentioned in Bourjade's memoirs, where he says that he coped with sarcasm by behaving in a friendly and exemplary manner. His rank as an officer also helped a little... Directly under his orders was his mechanic, Private Charles Maheu, who like many cases was to show absolute loyalty to "his" pilot even though he did not share his religious faith. The latter would give him a hard time with the arrival of his first aircraft, a Nieuport 27 fighter (n°5671).

As he wrote in his diary, Bourjade wished to place his aircraft under the protection of his patron saint to whom he wrote: "*Here I am at the front, and I have not yet offered you all my flights, all the battles that my duty will require of me, I have already, it is true, offered them to God and to you in my prayers from the first day, and I am full of confidence in your protection, all good, and your prayers, all powerful, on the heart of our Good Mother and of our beloved Jesus. I recommend to you at this time two very dear desires which are the object of the novena which I began on the 22nd. The first is that you help me in everything I must do for the vocation of the young man I wish to send to the little work (Apostolic School), and that you obtain for him the great grace of this vocation that you dispose his mother favorably. The second is that I be allowed to keep your portrait on the outside of the fuselage of my Nieuport, even though it seems that the regulations are against it. Goodbye, dear little Therese: teach me to love the good Lord!*"

He then wrote to the religious institution of Lisieux to buy a portrait of Saint Therese that his mechanic Maheu had to spread between a sheet of aluminum covered with a sheet of mica, to fix the whole on the fuselage near the cockpit. This kind of manifestation of faith was forbidden by the military authority whose generals had to apply the government's orders to enforce the principle of neutrality of the state. It seems that the installation of the portrait on the Nieuport took place during a period of uncertainty in the command of the squadron... Because on November 5, 1917, Lieutenant Charles Lefèvre, who was ill, was hospitalized. He was succeeded by another pilot of the unit, Lt. Lafargue, who was replaced on November 20 by a new officer sent by the general staff, Lieutenant Louis Delrieu. He was an experienced pilot, a veteran of the Eastern Front, and a future pilot for the Latécoère lines. He had no particular sympathy for Bourjade and his Catholic convictions, as can be read in the latter's correspondence: "*I am going to find myself the only officer with this Lieutenant Delrieu, who is not a very nice person to deal with, but this detail does not cause me too much trouble, on the contrary, it will give me a few opportunities to practice my patience.*" Bourjade met him on his way back from a leave that was granted to him from November 17 to December 3, 1917. It was from this date that he made his first flights in contact with the enemy, in a polar climate over the Vosges front.

Above: Léon Bourjade in front of his personal SPAD decorated with the portrait of Holy Thérèse.

First fights

The Nieuport 24, 24bis, and 27 that equipped the squadron were outmatched by the German fighters; in fact, the aircraft only equipped second zone squadrons in sectors that were not considered to be a priority, such as the Vosges front where the N 152 was located. Bourjade was judged by his superiors as a pilot without any particular qualities. Major Voisin, head of aeronautics for the 7th Army, wrote the following about him

Above: Léon Bourjade at the controls of another SPAD XIII, coded 13, probably an occasional mount. (Coll Albin Denis)

on December 5, 1917: "*Serious officer, average pilot. Has good will, trains rather slowly. Will get there if he continues to make a good patrol leader.*"

That same day he would encounter the enemy for the first time aboard his Nieuport, where he shot down a German plane that was turning back. He tells his mother about the fight in these terms: "*It was not a radical victory, but I did force my adversary to turn around and return home with a few bullets in his wings. As far as I am concerned, my aircraft received only one bullet which did not worry me at all*".

On February 20, he discovered a new target, a German observation balloon nicknamed "Drachen" or "sausage" by the French pilots. It was an easy prey, because it was inert and filled with hydrogen, which ignited on contact with incendiary bullets. Easy, but very dangerous to attack, because it was always protected by several pieces of flak and even watched over by fighter planes. The attack that he led on 20 February was not successful.

From December 29 to 31, 1917, he was detached to Luxeuil to reinforce the local aviation. Back at his unit, in January he had the flame of the Sacred Heart fixed on his Nieuport by his mechanic Maheu between two mica plates, attached to a stay just behind the cockpit. Nevertheless, Lt. Delrieu recognized the value of his pilot by supporting his request for a permanent appointment to the rank of second lieutenant, noting as an opinion "*Transmitted with a very favorable opinion concerning the appreciation of Mr. S/Lt Bourjade, from the point of view of his attitude in the squadron, where his dress as well as his way of serving are irreproachable in all respects.*"

Continuing his flights, he had another run-in with an enemy aircraft on February 10, 1918, claiming an aerial victory against an aircraft which was not granted. Seconded to Corcieux from February 13 to 15, 1918, when his squadron left this airfield for another one near Belfort (where Bourjade was housed in the local church rectory...), he had other unsuccessful fights, falling victim to machine-gun jamming. The cold was polar and his mechanic Maheu tinkered with an electric heating system connected to his gloves and shoes...

First victory

After going on leave from February 25 to March 12, 1918, he won his first air victory on March 27, 1918, against a Drachen whose attack technique he had mastered. He confided in his correspondence that:

"*Two years in the trenches as a crapouillot operator had left me with such unpleasant memories of the Drachen, that the day I saw my first victim collapse in flames, it was better than a victory, a revenge! Three times in a few days, attracted by this big game, inexperience, and also stage fright, had made me fail. I started shooting from too far away and my machine gun jammed. Although a little chilled, I was only waiting for an opportunity to start again. It soon came. That day, five Nieuports were patrolling from Munster to Thann. A beautiful balloon was keeping*

Above: Photo taken after the armistice in Haguenau, Alsace, in early 1919. From left to right we see S/Lt Ernest Maunoury, Lt Léon Bourjade, Lt Alexandre Sapiera (Polish officer), Capt Léon Bonne (squadron leader) and Lt Pierre Valluy (coll. Albin Denis)

watch down below, quite far away, about ten kilometers. I was at the back of the group and was carefully considering my attack. We had already passed the sausage. The observer, for a moment on the alert, had to resume his adjustment or his surveillance of the trenches. Without detaching myself too much from the others, I remained behind, I obliquely obeyed, then suddenly, I plunged, I pricked, I fell straight on my prey. It is a hard minute. I feel violent shocks in my chest, I do not hear any more my engine, but, only, the whistling of the fall. To the right and to the left, I hastily observe my wings: nothing abnormal. The rev counter? Oh, the breakup rpm. Quickly, I hit the engine control. The altimeter descends, but too slowly for my liking. And the sausage? It's there, well below. I approach it. It's growing very fast. Watch out! Damn! My machine gun is not engaged. Wham! That's it with a big punch. Ta ca ta ca, ta ca ta ca... Without aiming, I shoot. The belly decorated with its black cross, the enormous skin absorbs my puke at close range. Now, turn around, and in full sauce... But let's stay calm. Too much haste could lose everything. With all the strength of my senses, I observe my machine. The engine is humming like crazy. On the dials, everything is fine. The altimeter calms me down. I am at 700m. What about her? Missed? However I almost fell in! But no. It is there! What a joy... A small red point shows itself which, very quickly, grows, and all collapses in flames...

I run away. My glance searches everywhere, above, in front, behind, on the right, on the left. Not a single black spot. The luminous bullets of the German machine guns are now rounding their trajectories too low. A few shells crackle, but the pointer is surely not an ace. It is equal, the return is rudely longer than the first one! This time, however, everything went well. It was not always the same. For the Drachen hunter, this is the hardest time. Many people had to swear that they would never be caught again."

The victory was certified by a citation in the army order dated April 3, 1918, the same day he shot down a second balloon under similar circumstances, with the help of a crewman, by diving into the flak barrage: "*I saw the jets of bullets sometimes to my right, sometimes to my left, above and below, add to the party the cannon shots... So I confess that I was squeezing my elbows a little to make myself smaller... But Sister Thérèse flies with me.*"

Since March 29, 1918, Bourjade had been the interim leader of the N 152 squadron during a leave of absence of Lt. Delrieu. It was during this period that N 152 received a visit from S/Lt. Fonck, the ace of French aviation aces, who had come on his SPAD to visit his family in the Vosges. Bourjade received him and invited him to the squadron's table. Mechanics and pilots alike admired the stories of his battles...

Delrieu took over his command on April 8. A few days later, on the 12th, Bourjade received a new citation for his second success. Emboldened by his aerial victories, he noted in his correspondence (in a letter dated April 21) that: "*Now, no one will object to wearing my Sacred Heart flame.*"

The accounting book of the N 152 indicates that he was detached to the Villers-en-Lieu airfield from April 25 to 26, then to Manoncourt-en-Vermois from May 3 to 4. It was on May 4 that he burned his 3rd Drachen, followed by another on May 11, 1918. He was seconded to Dugny from May 15 to 19, and from May 22 to June 12, 1918, he went to the Cazaux shooting school.

Ace of the communiqué

This training course would be very profitable for him, since the aerial victories against the Drachen would from now on follow one another. When he returned to his squadron, it was now based in the Marne region of France at the Noblette airfield, near Châlons-en-Champagne, and was now equipped with powerful SPAD XIIIs. Bourjade immediately decorated his new aircraft (n°2262) with the portrait of Saint Theresa and the Sacred Heart banner. Promoted to second lieutenant on June 24, 1918, the next day he shot down his 5th Drachen, which made him an unofficial aviation ace, but from which he did not derive any glory, since the threshold for having his name mentioned in the military press release had been raised to 10 victories by that time.

However, he soon reached this threshold. On June 28, he shot down his 6th Drachen, followed the next day by a Fokker D VII which fell near Reims. Three other Drachen were burned by his machine gun fire on June 30, July 5 and 8, followed by three more on July 15, another on July 17, and another (his 15th) on July 19, 1918. Unfortunately, he was wounded during this attack by a bullet to the left shoulder and had to be evacuated to hospital n°57 in Vitry-le-François where his convalescence would keep him away from the front for six weeks.

It was there that the laurels of glory caught up with him. Already reported on July 15, 1918 for his "triple", on July 23, 1918, the communiqué to the armies mentions that "*Second Lieutenant Bourjade set fire to four captive balloons, three of them on July 15 and the fourth on July 17, which brings to thirteen the number of aircraft he has destroyed to date.*" If the total of 13 is inaccurate (in fact, he had 14 victories on July 17), the main thing is that this information was taken up in the national newspapers where his name, and even his picture, started to appear. Especially in the newspaper "*La Croix*", to which his parents were faithful subscribers, and which mentioned his status as a seminarian. He was given the nickname of "*l'as abbé*" (the priest ace) – publications mentioned that he blessed his enemies before attacking them, and that he chose the Drachen so as not to kill his enemies. This is an assertion that he wishes to deny, writing to the aeronautical journalist Jacques Mortane that "*Several times, in the newspapers, an interpretation of my feelings appeared that surprised me. I have never avoided the slightest combat against the Boche, whether in the trenches or in the air, in order not to shed enemy blood: what a singular idea of duty was lent to me there!* "His fame was there and from then on the newspapers would regularly report on his future victories.

Return to combat

On August 27, 1918, he returned to his squadron with lieutenant's stripes that he had received on July 31 during his convalescence. The SPA 152 squadron was now commanded by Captain Léon Bonne, who did not seem to create any difficulties for Bourjade and his religious convictions. These were reinforced by the arrival in the unit of his friend Henri Antoine Garin, a missionary of the Sacred Heart whom he had met in Switzerland, and who, like him, had started the war in the infantry where he had been decorated several times before becoming a second lieutenant pilot.

The two friends quickly teamed up in the air, even though Bourjade won most of his next victories with Second Lieutenant Ernest Maunoury. He burned 7 new Drachen with this pilot until October 1st (bringing his score to 22), then flew with Garin and shot down five new balloons during the rest of October. Unfortunately, Garin did not see the armistice, as he was killed in action on October 29, which had a serious impact on Léon Bourjade, who won his 27th and last victory against a Drachen on the same day.

This figure of 27 victories corresponds to those claimed on his own flight log. His last citation in the order of the day of January 28, 1919, recognized his aerial victory of October 8, 1918 as his 24th. Bourjade requested to be named officer of the Legion of Honor for his "25th and 26th victories" of October 27 and 29; however, this total lacks a victory won on July 19, 1918, mentioned in a citation to the Army Order of August 8, 1918, but which was forgotten in the count, i.e. a total of 27 certified victories. After the armistice, the press announced a total of 28 successes, but his file does not mention any late confirmations. Anyway, out of these 27 victories, 26 were against Drachen, which makes him the ace of aces of this specialty, slightly surpassing the ace Michel Coiffard who was shot down shortly before the armistice with 34 victories including 24 balloons. Bourjade flew a total of 245 hours and 45 minutes of wartime flights.

Remaining in his squadron after the end of the fighting, Bourjade was granted several long leaves from November 15 to December 1, then from December 13 to January 12, 1919 and from March 7 to 19, 1919, during which he visited his family. On April 4, 1919, he was demobilized and removed from the control of his unit.

Missionary in New Guinea

Back in civilian life, he went to Friboug in Switzerland to complete his training as a priest. He was not insensitive to military honors since on July 14, 1920 he wrote to the colonel of the 59th artillery regiment, his unit, to ask him to support his nomination to the rank of officer of the Legion of Honor, pointing out that his last aerial victories had not been rewarded. He obtained satisfaction on December 4, 1920 and the precious ribbon was given to him in Switzerland by the French

Above: Portrait of Léon Bourjade made in August-September 1918, when his war cross was decorated with 8 palms and a star.

consular authorities.

Ordained a priest on July 26, 1921, he embarked the following November for New Guinea to become a missionary, the dream of his entire life. Settling on the mission of Yule Island, about sixty kilometers north of Port Moresby, he traveled to various points of New Guinea for his evangelizing mission. There he discovered a situation somewhat different from what he had read in his youth. If the Roro natives were not the cannibals of his imagination, the tropical climate and the fevers that it conveyed were indeed there. Boujade gave himself to the task with ardor and was described by his fellow missionaries as having a reserved character. The Australian press reported the arrival of this atypical missionary who received a donation of 500 francs from Marshal Foch to build a church in Waima, in the interior of New Guinea. As a last reminder of his past, he agreed to wear his medals during a commemoration of the armistice organized locally by Australian settlers.

But during the summer of 1924, Father Léon Bourjade was affected by tropical fevers. His health deteriorated while he was in a coastal village building a bridge. Brought back to Yule Island, he died of hematuria on October 22, 1924, declaring that he was leaving in peace for the other world by receiving extreme unction. The Australian authorities paid him military

Above: Léon Bourjade, missionary in Papua New Guinea.

honors and the local press reported the loss of the "good and brave Bourjade".

In France, the news of his death was the subject of only one entry in the newspaper "*La Croix*" on October 28, 1924. The announcement of the mass in his honor at the Invalides, in the presence of General Gouraud and many of his former comrades, received a wider echo in the national press. The city of Montauban paid homage to him by erecting a monument to his memory on March 23, 1935, the Catholic world took note of his life, and two biographies were published on his life during the interwar period, an air scout squadron being named in his honor.

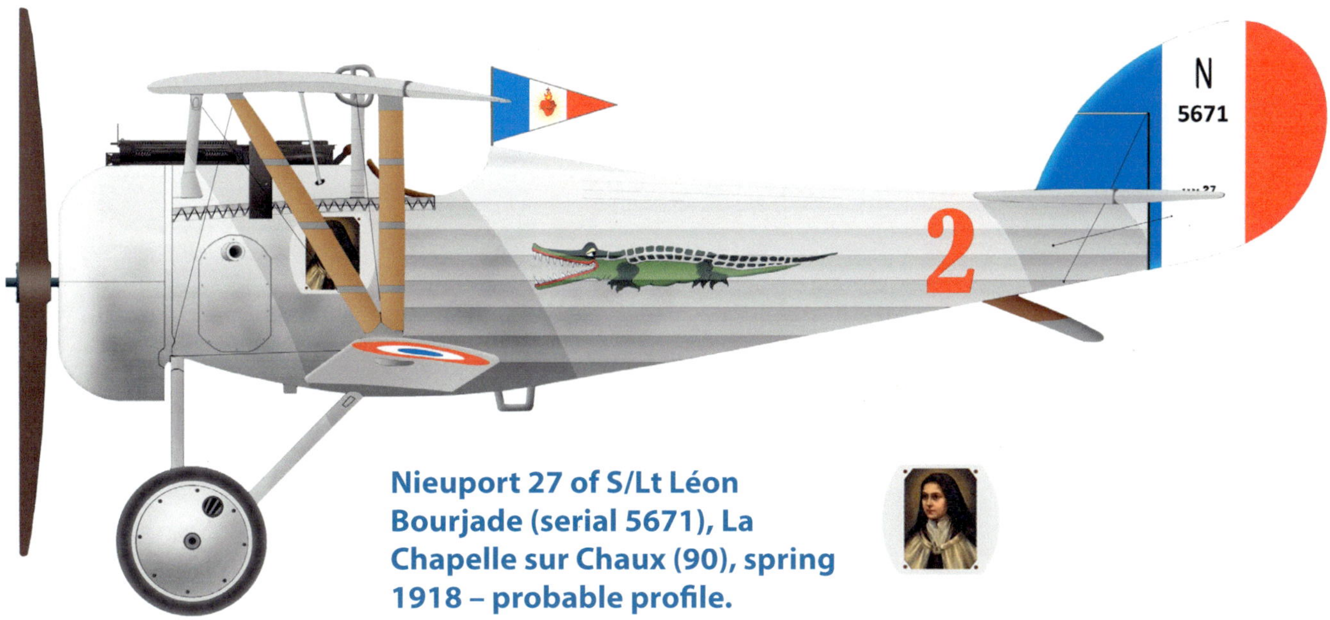

Nieuport 27 of S/Lt Léon Bourjade (serial 5671), La Chapelle sur Chaux (90), spring 1918 – probable profile.

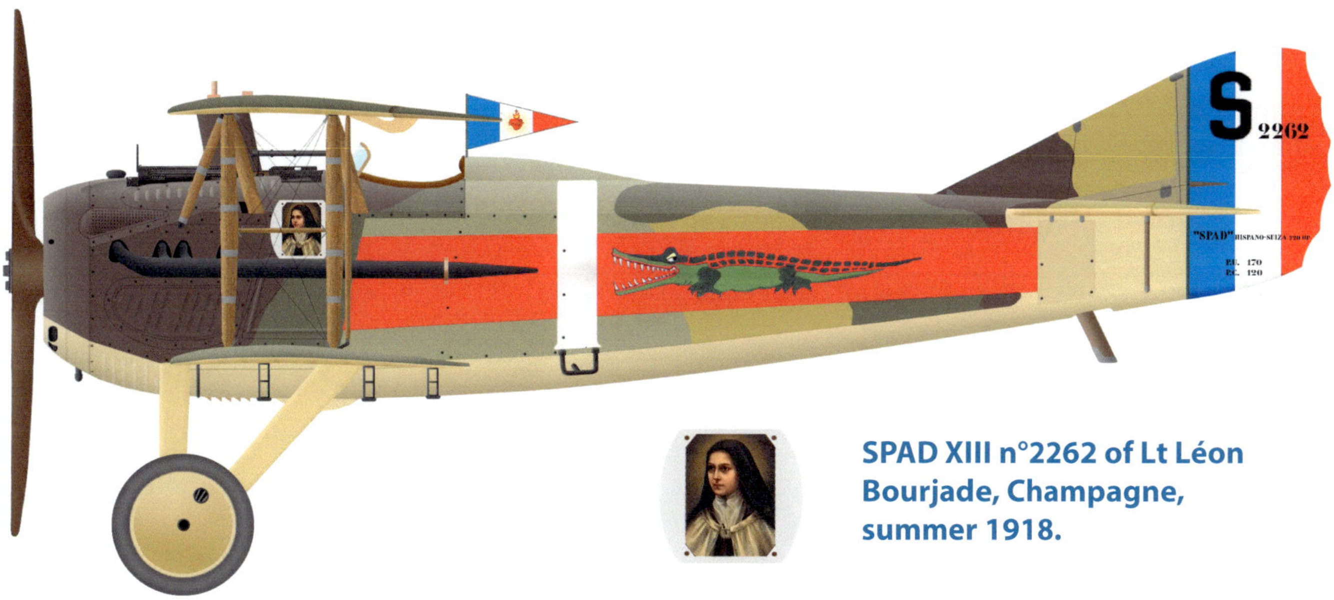

SPAD XIII n°2262 of Lt Léon Bourjade, Champagne, summer 1918.

Léon Bourjade Victory List							
N°	Date	Unit	Plane piloted	Plane claimed	Place	Notes	Confirmation
P1	2/10/18	N 152	Nieuport 27 n°5671	enemy plane	Ban-de-Sapt		
1	3/27/18	N 152	Nieuport 27 n°5671	Drachen	Geberschweiler		Citation OA 03-04-1918 - JORF 02-05-1918 p.3825
2	4/3/18	N 152	Nieuport 27	Drachen	Willer / Altkirch	With Brig Freslon	Citation OA 12-04-1918 - JORF 13-05-1918 p.4159 "flamed his second enemy device"
3	5/4/18	N 152	Nieuport 27	Drachen	Geberschweiler		Citation OA 27-05-1918 - JORF 01-07-1918 p.5716 "3rd enemy device"
4	5/11/18	N 152	Nieuport 27	Drachen	Willer	With Cprl Vallot, Cprl Bebely and Cprl Marsaudon	Citation OA 05-06-1918 - JORF 11-07-1918 p.5965 "4th enemy device shot down". Also Citation OA (LH) - JORF 30-07-1918 p.6610
5	6/25/18	SPA 152	SPAD XIII n°2262	Drachen	St. Etienne-a-Arnes		
6	6/28/18	SPA 152	SPAD XIII	Drachen	St. Pierre-a-Arnes		
7	6/29/18	SPA 152	SPAD XIII	Pfalz D.XII	N. Rosnay (Braucourt)		Citation OA 04-07-1918 - JORF 04-12-1918 p.10440 "5th to 7th victories"
8	6/30/18	SPA 152	SPAD XIII	Drachen	Marvaux-Rosières		
9	7/5/18	SPA 152	SPAD XIII	Drachen	Hauvine / Montfaucon	With Sgt Robert and Cprl Lac	
10	7/8/18	SPA 152	SPAD XIII	Drachen	Ivoiry / Montfaucon		
11	7/15/18	SPA 152	SPAD XIII	Drachen	Hauviné		
12	7/15/18	SPA 152	SPAD XIII	Drachen	Hauviné		Citation OA 15-11-1918 – "11th enemy plane" (WRONG, 12th)
13	7/15/18	SPA 152	SPAD XIII	Drachen	Main de Massiges		Citation OA 26-07-1918 - JORF 24-09-1918 p.8326 "in eight days, shot down 6 enemy Drachens (8th to 13th victory)"
14	7/17/18	SPA 152	SPAD XIII	Drachen	N. Nauroy		Citation OA 08-08-1918 - JORF 01-10-1918 p.8571 "flamed an enemy Drachen (14th victory)"
P2	7/17/18	SPA 152	SPAD XIII	Drachen	NO. Vaudesincourt		
15	7/19/18	SPA 152	SPAD XIII	Drachen	Tahure		Citation OA 08-08-1918 - JORF 01-10-1918 p.8571 but forgotten in the count of victories.
16	8/30/18	SPA 152	SPAD XIII	Drachen	Chavonne	With S/ Lt Ernest Maunoury	Citation OA 20-09-1918 - JORF 23-12-1918 p.11040 "15th victory" (wrong, 16th)
German casualties: Victory 7: Jasta 81, Gefr. Friedrich Bär, killed near Reims.							

Léon Bourjade Victory List							
N°	Date	Unit	Plane piloted	Plane claimed	Place	Notes	Confirmation
17	9/1/18	SPA 152	SPAD XIII	Drachen	Pont Arcy	With S/ Lt Ernest Maunoury	
18	9/4/18	SPA 152	SPAD XIII	Drachen	Pont Arcy	With S/ Lt Ernest Maunoury and Cprl Manson	Citation OA 02-10-1918 "16th and 17th victory" (wrong, 17th-18th)
19	9/15/18	SPA 152	SPAD XIII	Drachen	Ailles (Paucy)	With S/ Lt Ernest Maunoury	
20	9/15/18	SPA 152	SPAD XIII	Drachen	Hubertise (Paucy)	With S/ Lt Ernest Maunoury	
P3	9/29/18	SPA 152	SPAD XIII	Drachen	Hermonville		
21	10/1/18	SPA 152	SPAD XIII	Drachen	Orainville	With S/ Lt Ernest Maunoury	
22	10/1/18	SPA 152	SPAD XIII	Drachen	Orainville	With S/ Lt Ernest Maunoury	Citation OA 27-10-1918 - JORF 19-02-1919 p.1850 "18th to 21th victories" (wrong, 19-22)
23	10/3/18	SPA 152	SPAD XIII	Drachen	Cherisy-Cheret	With S/Lt Garin	
24	10/4/18	SPA 152	SPAD XIII	Drachen	Orgeval	With Adj Manson	Citation OA 25-01-1919 "21th and 22th victories" (wrong, 22-23)
25	10/8/18	SPA 152	SPAD XIII	Drachen	Chamonville	With S/Lt Garin	Citation OA 28-01-1919 - JORF 26-05-1919 p.5429 "24th victory" (wrong, 25th)
P4	10/27/18	SPA 152	SPAD XIII	Drachen	Ville-au-Bois	With S/Lt Garin and Sgt Lac	
26	10/27/18	SPA 152	SPAD XIII	Drachen	Château-Porcien	With S/Lt Garin	"My 25th and 26th victories" (personnal letter written on July 14, 1920, from Switzerland)
27	10/29/18	SPA 152	SPAD XIII	Drachen	Seraincourt		"My 25th and 26th victories" (personnal letter written on July 14, 1920, from Switzerland)

Jean Bouyer

First names: Jean, Georges.
Born: 12-07-1891 in La Rochelle (Charente-Maritime).
Father's profession: Delicatessen.
Military pilot brevet: 25-03-1917 (n°5769).
Units: V 396, N 49.
Army communiqué: 25-07-1918 (for 10 victories).
Total score: 12 victories, 1 probable.
Died: 16-01-1921 in Mourmelon-le-Grand (Marne) - killed in a flying accident.

Above: Post-war portrait of Jean Bouyer.

A mechanic from La Rochelle mobilized in the air force

Jean Georges Bouyer was born in La Rochelle on July 12, 1891 into a family of small businessmen, where his father was a pork butcher. The young man, rather tall for his time (1.76 m), lost his parents during his adolescence and had to quickly enter active life. He became a mechanic and had to do his military service in 1911, the year he turned 20. But, suffering from testicular ectopia, he was exempted.

He was called up for service shortly after the declaration of war, as a private 2nd class in the 114th infantry regiment. But he never saw the trenches: his physical inaptitude and his skills as a mechanic immediately made him suitable for the aeronautical service, where he was seconded on December 28, 1914.

He joined the C 30 squadron as a mechanic on January 25, 1915 and served in this unit for eleven months, during which time he was administratively attached to the 238th infantry regiment on April 12. Volunteering to become a navigator, he joined the machine gunner school of Chalais-Meudon on December 25, 1915, then was sent to the Cazaux school on January 3, 1916 for a shooting course. At the end of this training, he joined the V 396 squadron (later VC 468) on Voisin-Canon on February 19, 1916, a squadron of the entrenched camp of Paris based at Le Bourget and in charge of defending the capital from enemy air raids. The latter hardly appeared during the year 1917...

Jean Bouyer was probably bored with these patrols in a sky empty of enemies and volunteered to become a pilot himself. His request was accepted on December 16, 1916, when he joined the Dijon school for his theoretical training. He then went to the Crotoy flying school in the Somme estuary on February 1, 1917, where he obtained his military pilot's license n°5769 on March 25, 1917 on a Caudron G 3. Two days later, he left for the Avord training school where he was promoted to corporal on April 2, 1917, and then went to the Pau acrobatic school on April 25. Once his training was completed, on May 1 he was transferred to the Groupement des Divisions d'Entrainement (GDE), awaiting his assignment to an operational unit.

The ace of Belfort

On May 15, 1917, he was quickly assigned to the N 49 squadron stationed at Sermamagny near Belfort, where he remained until the Armistice. The sector of the front was relatively calm, nevertheless some aerial fights took place against the German planes stationed in Habsheim and other grounds in Alsace.

Jean Bouyer quickly distinguished himself there, perhaps thanks to his good shooting skills gained with his year of experience as a machine gunner. On June 25, 1917, he shot down his first enemy plane, an LVG C, and was promoted to sergeant and received his first citation in the army order on July 10. But even before receiving his Croix de Guerre with palm,

Above: Jean Bouyer among the wreck on one of his victims.

he scored a second victory on July 7, 1917 against a plane that fell between Suarce and Réchézy. This victory was rewarded with a citation in the army order.

It is thus with a war cross decorated with a palm and a vermeil star that he left on leave to La Rochelle on July 31, 1917, to meet his sister and his aunt. However, he suffered serious health problems and had to be treated at the hospital in La Rochelle from August 14 to October 20, 1917, thus remaining away from the front for two months.

He then immediately returned to his squadron and showed that he had lost none of his skill. In December 1917, he shot down no less than four aircraft, all of which were certified. The first one by a citation in the army order dated December 25, 1917, which specified that it was his 3rd victory, and all the others in the citation awarding him the military medal on January 4, 1918, specifying that he shot down 4 planes in December.

This detail is important when counting his successes, which continued in the Belfort sector during 1918, where N 49 was stationed, renamed SPA 49 because it was now equipped with SPAD fighters. On 19 February, it won a new victory, certified by a citation, and then on 18 May another one, certified by a citation that stated that this was its 7th success. However, if we go back to the count, it is his 8th – an error that will continue

Above: Portrait of Jean Bouyer.

in his other citations.

Flying with MdL Paul Hamot, his favorite teammate, he won another success on July 8 and a "double" on the 19th, then, promoted to the rank of second lieutenant on a temporary basis on September 13, he closed his record on October 21 at the expense of a two-seater shot down with Hamot and two other pilots, two other aces of the SPA 49, S/Lt Roger Brétillon and Adj Maurice Arnoux.

With 12 confirmed successes (and not 11, as indicated by error), he was an ace of aviation and saw his name published in the communiqué to the armies of July 25, 1918 on the occasion of his 10th victory, obtained on the occasion of the double of July 19. The communiqué was published in

Above: Jean Bouyer during winter 1917–1918 in front of his SPAD XIII near Belfort.

the national newspapers, where he was already mentioned before this date thanks to the articles of the journalist Jacques Mortane who mentioned him when he obtained his 5th success.

A short post-war period

After the armistice, he thought of joining the active army and made a request to that effect, supported by his superiors who emphasized his brilliant record of service and his sense of command. But this request was not successful: although he was confirmed as a reserve second lieutenant, he was demobilized on April 10, 1919 after having passed through the Dijon center.

Unemployed, he thought he would find a job in civil aviation. At the beginning of 1920 he went to Spain with his friend Marcel Aurouze to perform aerial exhibitions in San Sebastian on a Caudron G 3, owned by the latter. The business was not profitable and Bouyer returned to France. He found a job as a receptionist pilot for the Hanriot company.

Above: Jean Bouyer in a Hanriot HD 1 during the air meeting of Buc in 1920.

Jean Bouyer Victory List							
N°	Date	Unit	Plane claimed	Place	Notes	German Casualties	Confirmation
1	6/25/17	N 49	LVG C	Friesen			Citation OA 10-07-1917 - JORF 30-10-1917 p.8689
2	7/7/17	N 49	Enemy plane	Suarce / Réchezy			Citation OCA 31-07-1917
3	12/1/17	SPA 49	Enemy plane				Citation OA 25-12-1917 - JORF 22-04-1918 p.3508 "3rd enemy plane"
4	12/7/17	SPA 49	Enemy plane	E. Burnhaupt			Citation OA (MM) 04-01-1918 – JORF 21-02-1918 p.1757 "From the 1st to the 13th of december, shot down 4 enemy planes"
5	12/9/17	SPA 49	Two-seater	Masseveaux / Belfort	With S/Lt Henri Hirsh (N 151)	FA 289b, crew killed.	Citation OA (MM) 04-01-1918 – JORF 21-02-1918 p.1757 "From the 1st to the 13th of december, shot down 4 enemy planes"
6	12/13/17	SPA 49	Rumpler C (n°8273)	Montbéliard		FA 289b, crew killed.	Citation OA (MM) 04-01-1918 – JORF 21-02-1918 p.1757 "From the 1st to the 13th of december, shot down 4 enemy planes"
7	2/19/18	SPA 49	Fighter	Pfetterhausen			Citation OA 20-03-1918 "shot down one enemy single-seater that crashed on his lines"
P1	4/25/18	SPA 49	Enemy plane	Carpach	With MdL Paul Hamot		
8	5/18/18	SPA 49	Enemy plane	Burnhaupt	With MdL Paul Hamot		Citation OA 02-06-1918 – "7th victory" (WRONG, 8th)
9	7/8/18	SPA 49	Rumpler C	Burnhaupt	With MdL Paul Hamot		LO 27-07-1918 – JORF 01-10-1918 p.8574 "8th victory" (WRONG, 9th)
10	7/19/18	SPA 49	Albatros D	Guerweiler	With MdL Paul Hamot	Jasta 71, Vfw Karl Schmelcher killed.	Citation OA 28-07-1918 – JORF 22-09-1918 p.8272 "9th and 10th adversary" (WRONG, 10th and 11th)
11	7/19/18	SPA 49	Albatros D	Sulz	With MdL Paul Hamot		Citation OA 28-07-1918 – JORF 22-09-1918 p.8272 "9th and 10th adversary" (WRONG, 10th and 11th)
12	10/21/18	SPA 49	Two-seater	Mulhouse	With S/Lt Roger Brétillon, Adj Maurice Arnoux and Adj Paul Hamot	Gefr Helmut Back (pilote) killed in the sector	Citation OA 15-11-1918 – "11th enemy plane" (WRONG, 12th)
Plane piloted column deleted because there was no data known.							

After getting married on October 5, 1920 at the town hall of the 18th district of Paris with Miss Hélène Diquéro, he would participate three days later in the company of the most famous pilots of the moment at the Buc meeting to defend the colors of his company.

Unfortunately, three months later, he was killed in an aerial accident at Mourmelon le Grand on January 16, 1921, while receiving an aircraft. Several articles in the national press reported his death, presenting him as a great ace of the war.

Maurice Boyau

Name: Boyau, then Johannès.
First names: Jean, Paul, Maurice.
Born: 08-05-1888 in Mustapha (Algeria).
Father's profession: Entrepreneur.
Military pilot brevet: 03-02-196 (n°2705).
Units: N 77.
Army communiqué: 11-10-1917 (for 10 victories).
Total score: 35 victories (21 Drachens), 2 probables.
Died: 16-09-1918 in Harville (Meuse) – killed in aerial combat.

Above: Boyau in his works, photographed during matches of rugby before the war. Paul Delestre Collection

An accomplished athlete

Jean Paul Maurice Boyau was born on the 8th May 1888 in the commune of Mustapha, in the southern suburbs of Algiers, in the home of his parents, the "*Villa des lilas*". His family was recently settled in Algeria: his father, Jean Boyau, was a 43-year-old entrepreneur from the Landes region while his mother, Marguerite Nougier, was a 26-year-old housewife.

His family quickly left Algeria when he was a baby and returned to France to settle in Brétigny, in the Paris region, where was born his brother Emile and where he spent his childhood and adolescence. Then in 1907 he moved with his parents who settled on the place of their origins in the Landes village of Saint-Paul, near Dax. The young man detonated by his size, exceptional for the time: 1m81 for 75 kg! He became passionate about sport as many Frenchmen: under Anglo-Saxon influence, sports practices developed and began to benefit from broad media coverage with a magazine such as "*La Vie au Grand Air*", in which wrote the journalist Jacques Mortane, reporting on all the national sports competitions and narrating the exploits of the french "sportsmen". Maurice Boyau had been trying his hand at several disciplines with a predilection for rugby, and would soon see his name in this magazine as in other press titles. In 1908 and 1909 he was captain of the Dax Rugby team, then on October 8, 1909, he was incorporated at the age of 21 at the 144th infantry regiment of Bordeaux as a private soldier of 2nd class, having benefited from a carry-over of one year to complete his studies. It is under the flag that the young man disputed the final of the French Rugby championship with his new club, the Bordeaux stadium, which beat by 14 to zero the sporting university club of France (SCUF) – Boyau himself scored a penalty.

Released from his military duties on 29 September 1911, the future ace would be able to devote himself to his passion for sport and practice boxing, jumping, running and cycling. But he would particularly shine in rugby, more exactly "football-rugby" according to the terminology of the time. Captain of

Above: Boyau in his works, photographed during matches of rugby before the war. Paul Delestre Collection

the Bordeaux stadium, he was quickly selected on the national team and participated in the tournament of the five nations taking place from January to April 1912. Let's say it frankly: Anglo-Saxon players in this time are clearly superior than those of the French championship! Having integrated the tournament in 1910, the French team is far from being at the level of the Anglo-Saxon clubs and suffers defeat on defeat. She won the "wooden spoon" of the last one in 1910 and narrowly avoided her in 1911 by beating Scotland 16-15. The first match of the 1912 tournament took place in Paris on January 1 at the Parc des Princes, against Ireland. Boyau is in 3rd row and sees his team lose 11 to 6. The defeat is crushing against Scotland in Edinburgh on January 20, where it suffers a 31 to 3 ... Then two more defeats against Wales in Newport (14- 8) and against England at Parc des Princes on April 8 (18-8) where Maurice Boyau succeeded in a transformation in each of these games. This does not prevent a new wooden spoon for the team of France ...

During the remainder of 1912, Boyau now plays at the club of Dax with which he wins the local championship of the Basque coast. On November 30, 1912, a decree authorizes him to change his name to Maurice Joannès for unknown reasons - a family secret that is unknown but which seems to go back to his grandfather because his mother and his cousins will wear the name of Joannès-Boyau. Several Joannès also appeared in the civil status of Mustapha at the end of the 19th century. However, Maurice Boyau remains for the press and the general public, and it is under this name that he participates in the French team during the tournament of the five nations of 1913. He does not take part in the opening match on the 1st January 1913 against Scotland, which will remain in the

Above: This Nieuport (n ° 1831) captured in German colors is the first loss of the N 77 squadron, Lt José Maria forced to land in the German lines after a fight against the aircraft of the Flieger Abteilung 65. The aircraft carries a splendid personal decoration, characteristic of the Nieuport 23 of the N 77 squadron during the winter 1916-1917. Albin Denis Collection

Above & Below: Lineup of Nieuport of the N 77 on their airfield of Manoncourt-en-Vermois near Toul during the winter 1916–1917. One can see that most are decorated personally. Paul Delestre Collection

annals of the tournament. Unlike the first edition of 1910 which took place with a certain indifference, the French general public had now taken a passion for rugby and came en masse to the Parc des Princes in Paris. If the French team opened the scoring and made an illusion during the first half, it was quickly overwhelmed by the Scots who won the match by 21 to 3. Mr. Baster, the referee, was whistled during the game: an angry crowd crossed the safety cords at the end of the match! The police had to intervene and evacuate him by car with difficulty ...

Groups of supporters were spreading in the capital on the Place de l'Etoile (center of Paris), the avenue Mozart and the boulevard Montaigne to conspire his name – The French police had to charge to disperse these hooligans! On January 25th, England defeated France in London by a 20–0 and it was only in the 3rd match against Wales, at the Parc des Princes, that Maurice Boyau made his appearance, this time as captain of the team. If the match was more balanced, the team of France lost again 11 to 8. Boyau was again the captain in the last match against Ireland which took place in Cork on March 24, and where the team of France was again severely beaten 24–0. The incidents of the match against Scotland made waves to the point that the latter refused to face France for the tournament of 1914, which lost all its other matches despite a selection almost entirely renewed in which Maurice Boyau no longer appeared. Unable to face Scotland, France is thus implicitly excluded from the tournament. But the arrival of the First World War quickly shifted this sporting quarrel to oblivion.

Above: Maurice Boyau sitting on the fuselage of his personal Nieuport 23, decorated with a dachshund - The ace had a dog named "Trumpet" in squadron. Paul Delestre Collection

Mobilized

Maurice Boyau was mobilized on 2 August 1914 and joined the 37th colonial infantry regiment in Bordeaux as a private soldier. This regiment departed for Toulon and then went on the front in the Vosges mountains, where it saw some clashes with the German troops in the south-west of Colmar. But Boyau did not stay in the Vosges forests because he was quickly transferred to the 8th escadron du train (a logistical unit) in Commercy, in the Meuse, on the following 26th of October. He spent exactly 13 months performing routine missions at the rear of the lines to conduct mules or horse-drawn carriages. This assignment was far from suitable for him and he obtained, after several requests, to pass in the aviation and joined the school of Buc on November 26, 1915 to take courses of piloting. He graduated as a pilot on 3 February 1916 in Pau (military brevet n°2705) and received the rank of corporal two days later. But his disappointment was great when he discovered that he would not be affected in a squadron on the front, but kept in Buc as a monitor! It was not until September 24, 1916, after eight months spent instructing the student pilots, that he learned that he would be able to join a fighting unit, the N 77, a fighter unit undergoing training. He joined it on 12 October 1916 in his base in Toul, in Meurthe and Moselle, facing the front of the Vosges which was a sector where the fighting was quite calm.

The squadron was composed of 9 pilots flying on Nieuport 17 and also including a Nieuport 12 two-seater for observation flights. Most of the pilots were beginners, except S/Lt Raymond Havet, a veteran of the French army in Greece at N 87 attached to the Serbian army, where he won a victory near Salonika in 1916. There were also the Sous-Lieutenant Marcel Hughes and Adjudant Jacques Ortoli who would become aces in other squadrons. It would also be the case of Sergeant Gilbert Sardier, an accomplished athlete practicing running with which Boyau will quickly become friends. The squadron commander was Lieutenant Jean de l'Hermite, also a veteran of the French army in Greece and former commander of N 87, who was inspired by his pseudonym to find the symbol of the new squadron, the cross of Jerusalem representing the crusade of the beggars led by the preacher Pierre l'Hermitte and which would be painted in a blue flag on the fuselage of the airplanes after several months, ending a time period where pilots had individual colorful paintings in their Nieuports.

The first clashes with German aviation began in October and resulted in a loss, that of Adjudant José Maria that an engine failure forced to land in enemy lines on the 23rd and face capture. Adjudant Yves Barbaza claimed on the 25th a two-seater that he forced to land southwest of Tain but which was not confirmed to him. It was not until November 7, 1916, that Sergeant Sardier and S/Lt Georgeot in their Nieuport 12 officially shot down a German plane which fell at Vieville en Haye. Maurice Boyau, inexperienced, learned his job as a pilot at the front and was detached from 12 December 1916 to 6 January 1917 in escadrille N 12 where he received the rank

Above: View of the terrain of Manoncourt-en-Vermois under the snow taken by the observer of the Nieuport 12 flight of the squadron, Sgt Paul Delestre. Paul Delestre Collection

of sergeant. The winter limited the flights and Boyau would benefit from a long leave from the 31 of January to the 16 of February 1917.

A phone call from Guynemer

Returning to his squadron, he resumed his patrols without success for nearly a month, until the day he made a mission with his teammate and friend Sardier who recounted the meeting: "One day being on patrol with the great Maurice Boyau, we attacked a German two-seater above Pont-a-Mousson. We were very well placed, above him, and we fired one after the other. It was a pilot who defended himself well and an observer also, who did not spare us. We spun around without the desired result, when suddenly a SPAD of which it was the beginning comes between us and the German plane, shoots a few bullets (we have seen a dozen bright). The German caught fire, cut himself in two, and fell. We were somewhat dazed when this French SPAD plane came to idle around us which had only Nieuport less fast. And we read on "Old Charles". It was Guynemer! We go back to the field and start explaining the fight. We could not get over this story when a mechanic ran to us and said, "Captain Guynemer is asking on the phone the two pilots who saw him in combat during the patrol." We rushed to the phone which was I remember at the bottom of the hangar. The great Boyau who was at least 1 meter 85 told me "Take it! Take the phone! He looked like a great communicant. The conversation was as follows:

- Here Captain Guynemer.
- Our respects, Captain.
- I apologize to you, but I saw that you could not get down this German plane, so I intervened.
- But we congratulate you very sincerely, you did well.
- Will you allow me to give you some advice?
- Of course!
- Well, you were too far. To get down a Boche, you have to get him in!

It was in 17 February 1917 (...). These words are textual, engraved in our memory! "

The most probable date indeed seems to be March 16, 1917, when Guynemer actually shot down two two-seaters in Lorraine. However, his advice was well integrated by Boyau, who shot down the same day his 1st victory, while with 3 comrades he escorted a Nieuport two-seater returning from a reconnaissance mission on Metz. He told the story to the journalist Jacques Mortane: "*I was in liaison with Lieutenant Havet. We were back. The forest of Mortmare was there almost beneath us. Suddenly I noticed the absence of Havet, who, as agreed, was to remain on my right. I look in front, behind, to the left, to the right, nothing! Then I distinguish towards the southwest some explosions. I think immediately that the lieutenant has set course in this direction and I go there full engine. A few moments later I see two black spots in front of me. They are two*

Above: Splendid autochrome showing another Nieuport of escadrille N 77 during winter 1916-1917. Albin Denis Collection

planes. I see them only imperfectly. Is it two French? Is it two Boches? I keep holding them in my line of sight. I'm 15 meters away and I see ... a Nieuport followed a few meters by a hunting Aviatik. Of course, the French pilot did not suspect anything and, in flight line, quietly, continues his route. Crazy at this thought, I shoot some machine gun bullets to warn him. Immediately, he executes a reversal and finds himself face to face with the German. The battle begins ... Before I could intervene, Lieutenant Havet, for it was him, was defeated. I dived on the victorious Boche and in a few seconds I made him pay the price of his victory. My teammate was avenged. "

The German was confirmed but the loss revealed by the German archives concerns a two-seater Aviatik, which fell near Thiaucourt, killing its crew, Uffz. Karl Kolrop and Ltn. Feodor Kellner of FlAbt 279 (A). As fighting was still rare in the Vosges sector, Boyau obtained to organize a raid against the enemy airfield in Marimois, near Thiaucourt, which he realized with Sergeant André Boillot, brother of the pre-war automobile champion fallen in air combat at Verdun in 1916. The two pilots take small bomb-shaped shells with fins under the wings of their aircraft, which they drop at 250 meters on enemy hangars left in flames ... This gave to Boyau his first mention in dispatch, which also signaled his first aerial victory - the ace now wears the *croix de guerre* with palm to his chest. Other palms will soon be added to the ribbon.

A hunt for sausages

His squadron left Toul on 23 March for Manoncourt en Vermois, 30 km east of Nancy, still on the front of the Vosges. Boyau left the squadron on 27 March for the general aviation reserve in the Paris region. This is where he participated in a rugby match in the France Military team, against the team from New Zealand. Formed of players hastily removed from the trenches, without physical preparation, the team of France was massacred 40–0 by the All Blacks. On April 14th, Boyau was granted a well-deserved leave on May 3rd and did not return to combat until May 19, 1917. The N 77 had meanwhile integrated new pilots, including Captain Joseph Batlle, officer in second. As captain de l'Hermitte flew very little, his deputy would ensure the bulk of the operational flights and under his de facto command would be initiating an attack against a new objective that would become the specialty of Maurice Boyau: the Drachens, captive observation balloons, sometimes dubbed "sausages" by French airmen. Batlle testifies:

"We usually left in patrols of 3 pilots. Only when we were attacking a Drachen would we increase the strength. There were two elements: the one who was chosen to take him down, and those who protected him. There was no maneuver: we knew the position of the Drachen. They were brought up in enemy territory, we knew them. It was always necessary to attack when the wind was contrary to the direction which it was necessary to use to go to the enemy, so that the wind brought it closer and inclined a little, and so that when it blazed the flames quickly gained the summit (...) When we attacked a Drachen, it was a decision that was taken to the ground. Someone said, "We're going to fuck a Drachen at a place, because he fucks us, he fucks artillery, or because he directs his shot at a certain

Above: Only experienced pilot of the N 77, Lt Raymond Havet, veteran of escadrille N 87 commanded by Cpt de l'Hermitte where he won a victory near Salonika in 1916. He will fall in battle on March 17, 1917 under the eyes of Maurice Boyau, who will defeat his winner and win his first confirmed victory. Paul Delestre Collection

Above: The N 77 is based at Manoncourt in Vermont (Meurthe et Moselle) on March 24, 1917. There is Maurice Boyau (the tall man, in the center) accompanied by Sgt Paul Delestre (cross) and Sgt André Boillot (left, in dark uniform, head turned) doing a little tourism in the village. Paul Delestre Collection

Above: Splendid view of the Silver Nieuport of adjudant Gaillard in flight. This photo shows that the aircraft of the squadron now carry the badge of the N 77, the blue flag decorated with the cross of Jerusalem instituted by the Cpt de l'Hermitte. Paul Delestre Collection

place." The pilot who blazed it was designated by the community, not by the squadron commander, by the whole! And the others were designated by the squadron leader who said, "You, you, will put yourself in such a position to protect him." It was the shield ... The one who was designated to flame, to attack, is the one who took the most risks."

It was also Batlle who institutes the prayer of pilots, recited with devotion before each dangerous mission: *"Good holy virgin, pray for the airmen on the ground. Those who are in the air, they are fools, let them break!"*

On May 23, 1917, Boyau attacked his first Drachen with his Nieuport equipped with incendiary bullets. The observer jumped in a parachute while the Drachen was brought back to the ground by a winch. The ace did only succeed in damaging it, without igniting it. It was not until June 3, 1917, in collaboration with Sardier, that he succeeded in igniting one in Geline at 9.55 and then in damaging another one whose observer jumped with a parachute.

Henceforth, the tactics were honed. But the second Drachen he shot with Sardier two days later on June 5 at Moussey did give him cold sweats. Boyau, Sardier, and Rebourg left for a photo reconnaissance mission on Sarrebourg. On their return, around 10:55, they saw a Drachen that Boyau attacked at once. His Nieuport dived while shooting the Drachen pulled back to the ground. The machine-guns on the ground retaliated and surrounded it dangerously with tracer bullets. But Boyau succeeded at the last minute to inflame its target while the observer evacuated in parachute: *"I went down from 4000 to 400 meters. When I wanted to recover the height to return, my engine, lack of pressure, did not want to resume. All my attempts to use the means of rescue used in such cases fail; and then I realize that my engine lacks fuel. I am only 200 meters away and, despairing of regaining the lines, I notice a meadow beside a road. "* Along this tree-

CAFÉ
EPICERIE MERCERIE
Manoncourt-en-Vermois

Adj.t Gaillard par Adj.t Rebourg - Mono. N.77. 21-5-17 - Alt. 2500m

Above: The squadron also received during this period its first SPAD VII. One can see here that of the MdL Jean Sardier (1897–1976), which ends the war with 15 confirmed victories. Paul Delestre Collection

lined road two German machine-gun cars arrived, who saw the French plane and approached it to capture it. But Boyau furiously operated his pump ... and the engine resumed! He then took off from the German soldiers' nose, and according to the story of his squadron comrade Henry Decoin, *"leans out of the cabin, turns in a curve, and with his fur-gloved hand gestures with all his might characterized by one of our most fiery academicians ...* 'If memory has forgotten the name and the exact gesture of this academician, one can only guess the one that Boyau addressed to the enemy... Maybe a middle finger in the air, knowing that he had to keep one hand to hold his broom handle! All of his comrades worried because Sardier said he lost sight of his team-mate on the ground. A telephone call from the concerned person at Luneville came to reassure them. His third confirmed victory earned him a new mention in dispatch and the médaille militaire.

Boyau reached the envied ace title on 24 June 1917 by scoring twice. Described in the attack patrol with Sgt. Boillot and Sous-Lieutenant Hautefeuille, he managed to ignite a Drachen between Nancy and Goin while the protective planes were attacked by a group of five German fighters. Boyau then entered the fray and managed to shoot one that falls between Goin and Chevisey. A new palm comes to adorn the ribbon of his cross of war, followed by another on 13 July 1917 when in his, along with the Sous-Lieutenant d'Hautefeuille, he shot down a LVG C from Flieger Abteilung 276 which fell near Nancy at 10.30 and constituted its 6th victory.

On SPAD

After going on an authorized leave on July 23, 1917, he returned to his squadron on the 9th of August to discover that some SPAD fighters had been delivered to the squadron now commanded by Captain Pierre Mouronval, as Lhermitte was sent to higher rank functions. The pilots of SPA 77 did immediately use their new fighters offensively by attacking enemy positions: on August 23, 1917, Boyau was part of a group of five (with d'Hautefeuille, Sardier, Rebourg, and Boillot) who, in the evening, set off hand-held bombs at the train station at Fresnes en Saulnois, where they blew up barracks and machine-gunned the troops. The next morning,

Above: Nieuport 23 crash. Paul Delestre Collection

Below & Below Right: Boyau discusses with his squadron comrades. In the picture at right Captain de l'Hermitte wears a colonial helmet, a reminder of his period of command of the N 87 squadron in Macedonia in 1915-1916. Paul Delestre Collection

Above: SPAD VII of the squadron. Paul Delestre Collection

Above: Boyau in his SPAD VII. The unusual notch around the cockpit, to allow the ace of 1.81 m to slip more easily ... Paul Delestre Collection

Right: Boyau prepares for a mission in his SPAD VII. ... Paul Delestre Collection

Above & Below: Sergeant Maurice Boyau receives the military medal after his exploit of June 5, 1917, when he broke down in the enemy lines after shooting down a drachen (his 3rd victory), he manages to repair and take off the nose of two enemy machine-guns to capture him. Note that from the top of his 1m81, it clearly surpasses a head the officers came to give him his decoration ... Paul Delestre Collection

Left: During the First World War, Japan went to war with the allies and immediately invaded the German possessions in China and the Pacific. It also sends ships fighting the U-Boats in the Mediterranean, as well as pilots on the western front scattered in the French squadrons. The N 77 will receive two, Lt Isakitchy Nagao (here in the picture, arrived June 6, 1917) and Corporal Tadao Tamanaka who accidentally died on 5 December 1917 at the helm of a SPAD XI two-seater. Paul Delestre Collection

Below: In this photo where every member of the squadron seems to laugh after a joke done by Maurice Boyau, we can see the two Japanese pilots (under red arrows), Corporal Tadao Tamanaka (left) and Lt Isakitchy Nagao (right). Paul Delestre Collection

Above: Photo of the pilots of escadrille N 77 in Manoncourt-en-Vermois, July 1917. From left to right : n° 1 Caporal Henri Guillaume n° 2 Sgt Maxime de Ginestet de de Puivert - n° 3 Lt André Georgeot - n° 4 Cprl Ignacio de la Torre - n° 5 Cpt Jean de l'Hermite (commanding officer) - n° 6 Cpt Pierre de Mouronval - n° 7 Sgt Maurice Boyau - n° 8 Unknown - n° 9 Sgt Yves Barbaza - n° 10 Caporal Pierre Berthier - n° 11 Sgt (gunner) Paul Delestre - n° 12 Sgt André Boillot - n° 13 Caporal André Géhin - n° 14 Adj Gilbert Sardier - n° 15 Adj Henri Rebourg"

Right: German plane shot down by escadrille N 77 and photographed by Sergeant Paul Delestre

Above: Captain Joseph Batlle (4 victories, including 1 at SPA 77), deputy captain of l'Hermitte (whom he replaced in most operational flights), then deputy captain Mouronval, will be transferred to the squadron SPA 103 on March 27, 1918, of which he will command until the end of the war. Paul Delestre Collection

the same pilots (less Hautefeuille) bombed the Metz-Frescaty airfield and then, in the afternoon, the station of Hampont. On August 26, 1917, Boyau set off again on his Nieuport to the Drachen hunt and missed the first two he attacked, but exploded the 3rd in Juvelize-Bourdonnaye at 15:00. His 7th official victory earned him the relative celebrity of an article appeared a few days later in the magazine "*La guerre aérienne illustrée*" of the former sports journalist Jacques Mortane, who greeted the ace as he greeted the sportsman a few years ago ... Maurice Boyau, at the controls of his Nieuport, will flame a new sausage on September 16 with the sous-lieutenant Rebourg, another on September 22 with the same crew member, then shots down a German two-seater above Coincourt the next day, reaching the threshold of 10 victories that did earn him the honors of the army communiqué.

Glory in the sportsmen's squadron

The SPAD have now definitively replaced the Nieuports and it is on this fighter that Boyau won on October 1, 1917 his 11th victory against a two-seater from Flieger Abteilung 12 which fell in flames north of Champenoux at noon, in collaboration with Sous-Lieutenant d'Hautefeuille and Capitaine Tourangin from N 89. The honors of the release caught him on October 10 when he was promoted second lieutenant temporarily, then two days later the national press, delayed, echoed his 10th victory. He was appointed Knight of the Legion of Honor on 12 October and enjoyed a new leave from 15 November to 7 December 1917.

If winter limited the flights, Boyau increased his score on January 3 against a balloon flamed in Benay, then another flamed in Vaxy on February 20 with his friend Sardier. The army did not hesitate to show him at sporting events organized to maintain morale. Thus, while the accounting booklet of his squadron always mentioned him as present, he was therefore on an official mission when he took part in a football match on 5 February 1918 in Paris, where he took a team of youngsters against the team of 33rd Infantry Regiment. On 17 February he was still officially in his squadron and was the captain of the French Rugby team in a friendly match against New Zealand, including the biggest players of the day mobilized under the flags like Géo André or Philippe Struxiano. France lost at the end of a Homeric match, by the score of 5 to 3 much more balanced than that of the previous year.

SPA 77 left the front of the Vosges on March 8 to reach the Saponay airfield in the Aisne. Our ace officially enjoyed a new leave from March 30th to April 17th, and then performed two missions to the General Reserve of Aeronautics. Returning to his squadron, which was stationed near St Quentin on May 21,

Above: Maurice Boyau, face to the end of the wing of the SPAD, discusses the mission to come with another pilot while Sergeant André Boillot, brother of the famous racing driver killed in combat in 1916, adjusts his helmet to go on mission. Paul Delestre Collection

1918, Boyau saw two days later an old acquaintance joining his ranks: Adjudant-Chef Emile Strohl, rugby international player and who played with him the final of the Rugby championship of France in Bordeaux stadium in 1911, and the recent match against the New Zealanders a few months earlier. The SPA 77 squadron also includeed Lt. Henry Decoin, the 1911 French champion in swimming 500 meters, Sous-Lieutenant Yves Barbaza, a high-level swimmer, Sergeant Richard Mewius, tennis player, and Captain Pierre Mouronval who also practiced rugby. The press would soon call the formation "sportsmen's squadron", whose aura of ace Boyau did energize the fighting spirit. The German spring offensives of 1918 did give him the opportunity to exercise it.

Facing the German offensive

On May 29, 1918, SPA 77 settled for a day in the South-east of St. Quentin, before being ordered the day after to leave for Champaubert to 30 km to the South of Reims. However, it was on this day that Boyau, flying a cannon-SPAD XII given only to the aces, won a "double" by shooting down an Albatros DV with Sardier in Ville-en-Tardenois (plus a probable one), and a sausage on the Dole Forest in the southeast of Fismes: *" On arriving at the lines, I was sad to see one of our Breguets of reconnaissance attacked and defeated by two Albatrosses. I had no time to intervene, but I might have the one to avenge the unfortunate crew. I arrived. The Boches, who were to gargle their success, probably did not see me coming. I hastened to shoot: one collapsed in our lines. The other was running away. He had to catch up with him: it reminded me of my football games. In a few wings I joined him and made him share the fate of his comrade. Then I went to the sausages. Oddly enough that I did not explain myself yet, I did not put my target into flames, but I cut it in two as with a knife. One piece fell to the left, the other to the right and the nacelle did not seem to know which one to follow. It was during this hike that we had the pain of seeing our chief and revered friend, Captain Mouronval, disappear in the enemy lines where he was captured. The next day I was going to attack a sausage without success and our excellent comrade Gehin was missing in his turn. "*

Boyau did not know that Captain Mouronval was killed in battle. Lieutenant Henry Decoin succeeded him immediately: the losses of the squadron accumulated during these missions carried out at a steady pace and which included numerous attacks on the ground with machine guns against the German troops, which progress. But the victories of our ace also: a Pfalz D shot down on June 1, two sausages shot with Sardier on June 4, and another on June 27. He won his 20th confirmed victory on July 1 against a sausage shot down in collaboration with the ace Marcel Haegelen of the SPA 100, which is its eighth victory.

Above, Above Right, & Facing Page: Maurice Boyau visits the trenches on the Vosges front with several pilots from his squadron, including Sgt observer Paul Delestre. Paul Delestre Collection

July 5, 1918 is the occasion for Boyau to win another double, under circumstances reported by his squadron leader Henry Decoin: *"The day had been calm. The low clouds, very low, the sun was muffled, a strong wind from the west, all of this had allowed the mechanics to repair the zincs and the pilots to repair themselves. However, towards the evening, we were going to take a turn to the field, history to see. The sky always gray and the wind always strong! Boyau, the great Maurice, walks up and down, followed by his faithful dog Trumpet, low-red of pure race. Finally, without saying anything, he decided. His mechanics take out the 2-machine-guns-SPAD. He puts on his silk stocking which serves as a pass-mountain, his enormous blue sweater, color of the Racing and it interferes in the narrow cabin. The engine is running. The mechanics, with wings, place the bird in front of the wind and the plane, so many times victor, escapes from earthly embraces and climbs towards the gray and dirty vault. And over there, on the lines, this is what happens: Boyau, above R ..., sees in the distance a black spot. He believes in an enemy aircraft in flight, by this time it really is a boon. He waits a few seconds to see where the Boche is heading. But the black spot does not move! Then Boyau realizes that the black point in question is a superb sausage. Without hesitation and close to the clouds, he moves towards his prey. They are far enough, the Boches sausages: 8 to 10 kilometers behind their first lines. Obviously, to go there, we always go! But it is the return, which is sometimes quite painful ... The boche Drachen, now accuses its forms. It is only 300 meters away, 200 meters. The funny thing is that he has not moved yet. Boyau disappears into the clouds after properly spotting the drachen. He sits down, and suddenly dives. Its two machine guns work admirably. They shoot, but shoot ... It's a pleasure! And the Drachen, surprised, lets his observer escape with the aid of the parachute and ignites gently ... It is pretty as a whole boche Drachen burns! Boyau, in spite of the bullets of the machine-guns, the cannon-shot, the incendiary caterpillars, turns and looks at this fairy-tale spectacle! Only the Boches saw the blow and several aircraft rushed from a nearby airport to punish the French pilot. There are four of them. But of the four there is one that rises much faster. Boyau is waiting for it! The Boche arrived and rushed headlong. The combat takes place at 300 meters of altitude and always behind enemy lines, at 8 kilometers. A few quick passes and the aircraft and the pilot leave, always in flames, crashing near the Drachen which is consumed on the ground. The other three, at the sight of this kind of sport, go up less and less quickly. And as they really do not put any good will into it, Boyau is obliged to return to French lines. "*

The count of the victories is now at 22 and Boyau entered the restricted club of the very big French aces: an article in the biggest French newspaper *Le Petit Parisien*, signed by Jacques Mortane and dated 14 July 1918, signals to the public that he was now the 4th scoring ace fighting at the front, behind Fonck, Nungesser, and Madon.

The fighting continued and the Germans launched their last offensive in Champagne on 15 July, threatening the city of Reims. SPA 77 was in the front row and Boyau shot down a Fokker D VII that day, followed by another enemy aircraft on the 17th. Two days later, a French counter-offensive took back all the lost ground and now the German troops, having exhausted their last reserves, remained on the defensive during the rest of the war, yielding ground against the combined attacks of the French, British, and American troops. Boyau won his 25th confirmed victory on 21 July against Otto Esswein of Jasta 26, an ace with 12 victories who fell in Hartennes south of Soissons. The following day, 22 July, he won a triple victory by flaming two balloons with the adjudant Guerrier as wingman, then an enemy plane that fell in Fresnes. The pace of his victories went so fast that the press could no longer keep up, because the day's edition of the *Petit Parisien* celebrated Boyau as one of the biggest aces by announcing the Confirmation of his 24th victory ... While he had 4 more! As an officer of the legion of honor, the army used its hero by showing him in the rear – he assisted on August 5, 1918, with the ace of aces Rene Fonck, at the sporting meeting of Colombes in which is given to the public a demonstration of "assault artillery", tanks that cross an obstacle course.

Returning to the battle, he won a new triplet on August 8, 1918. He left in the late afternoon with teammate Maréchal des Logis Antoine Lentz-Mitchell. They aimed the Drachen of Cury that Boyau shot without hesitation while his teammate remained in protection then dived to accompany the return of its chief who crossed the lines at 200 meters of altitude. A German two-seater took the opportunity to slip behind Lentz-Mitchell and attack him. But Boyau saw it and sent it to the ground with a well-placed burst. The two SPADs resumed their route to the French lines, when a shell hit Lentz-Mitchell's plane and pulverized it in the air. Powerless, Boyau could only continue to his lines when he was in turn attacked by a two-seater, that he shot down near Ressons-on-Matz in the enemy lines. The two planes would be confirmed to him later, with the advance of the French troops, as his 30th and 31st victories.

He was again called to the rear and participated on August 18, 1918 in a military athletics competition at the Bergeyre stadium in Buttes-Chaumont and bringing together personalities from the sports world serving the armies, including athlete Geo André. He was also in good shape as evidenced by Sergeant Pierre Maria, one of the pilots of SPA 77: *"Near the mess of the officers there was a basin. With three other comrades, we decided to throw Boyau in it all dressed ...: Result we ended up all four in the water!"* Appointed on the same 18 August second lieutenant definitively, he could leave two days later for a final leave which ended on 12 September.

The month of September 1918 saw the German army collapse under the blows of the Allied armies. SPA 77 was now based in Rancourt in the Meuse and accompanied the French offensives. This was the month when victories rained for the French airmen. Scarcely had he returned from leave on 13 September 1918, Boyau left at 4 pm on a patrol in which he was attacked by a squadron of 6 German fighters. He managed to damage one, but was himself riddled with bullets and had to land in the first French lines and spent the night in a command post. It was a first warning of destiny ... Boyau did not stop and the next day he managed to burn a Drachen with Lt. Haegelen and Corporal Corsi. The following day, patrolling with Sous-Lieutenant Barbaza, Lieutenant Decoin, and Adjudant Strohl, he added two more balloons to his record.

The last mission

September 16, 1918 would be the day of his last mission. Maurice Boyau left late morning on his SPAD XIII n ° 10666 with the Aspirant Henri Cessieux to attack a Drachen signaled in Harville (Meuse), halfway between Verdun and Metz. The notebook of Sergeant Jean Moulis, young pilot at SPA 77, informs us about the events.

"Boyau and Cessieux will go on the attack of a drachen, and succeed in flaming it. On the way back, seized by seven

Above & Facing Page: The great Maurice Boyau in front of his SPAD XIII equipped with rockets Le Prieur, on the grounds of La Fère en Tardenois on 11 March 1918. Paul Delestre Collection

enemy fighters, Boyau did engage the fight, and remove his pursuers. Flying low, he did dive vertically, most likely shot by bullets shot from the ground. Cessieux saw him crash into flames, near the lakes of La Chaunié. We did never find the place of the accident nor the debris that could identify him. The enemy did then pursue the remaining French aircraft, and our friend, who was attacked in the back, did have the strength to land on the first lines of our home, he did jump out of his cabin and fainted. The artillerymen evacuated him. Meanwhile, the Germans, flying on the ground, tried to destroy the plane. Soon after, artillery shells did spray it between the lines. We were waiting impatiently on the ground for the return of our comrades. Alerted in the course of the day by the ambulance where Cessieux was evacuated, we learned with consternation

the end of our great ace. Mourning is cruel, but the task is not over. The next day, we resumed the command of our planes with the intention of avenging him. "

A further investigation into the first French lines would revealed that Boyau did indeed flare the Harville Drachen at 11:17. Returning to an altitude of 200 m, he was pursued by a Fokker D VII 500 m behind. Then his plane was seen to stick vertically and crash, probably hit by a shot of the ground. 3 Fokker D VII with a red nose, carrying a white Greek cross on the fuselage and on the tail unit, with two red and yellow oblique strips on the fuselage, flew over the crash site and rotated around for some time. The very young Leutnant Georg von Hantelmann (20 years old) claimed the plane of Boyau, shot down southwest of Conflans-en-Jarnisy at 11:20, which would be confirmed as his 14th victory of the 25 that he would win during the war. If the place, time, and even decoration of the Fokker of the Red Nose Jasta 15 plead in favor of von Hantelmann's claim, the distance of 500 meters behind which was his Fokker D VII makes his shot very random. It will never be known whether Boyau succumbed, as seems more likely, to a bullet fired from the ground or a lucky blow from the hunter's twin machine guns.

The press announced his disappearance on September 21, 1918. Some of his comrades still hope ... It was without his captain, dead in the field of honor, that the French military team of "football-rugby" confronted New Zealand in the Parc

Above: The great Maurice Boyau in front of his SPAD XIII equipped with rockets Le Prieur, on the grounds of La Fère en Tardenois on 11 March 1918. Paul Delestre Collection

des Princes. On November 4, 1918, an athletic competition that bears his name was organized in his honor and was won by a team composed of SPA 77 pilots, the "Sportsmen's Squadron", led by Henry Decoin. The "challenge Maurice Boyau" will be organized until the 1930s.

However, Maurice Boyau was still officially missing in action after the armistice and a curious story did keep open his file. Maurice's mother did everything she could to know his fate, asking the help of many French politicians. She was contacted by a Frenchman living in Holland, a certain Harry Mandelson, who claimed to have met in January 1919 a certain Leutnant Wendt repatriated from a prison camp in England. This officer, amputated with one arm, would have himself crossed Maurice Boyau in a health facility, which would have asked him to transmit to his family his military papers and his identity plate. The services of the French army conducted a thorough investigation and found no Leutnant Wendt corresponding to the report given to them. On the other hand, they soon discovered that Mandelson was a merchant of hope, a weekly swindler whose wife practiced prostitution, and who informed the German services during the war about deserters who had taken refuge in Holland. The Boyau case was closed on January 21, 1922 when the Civil Court of Dax rendered the declaration of death of Sous-Lieutenant Jean, Paul, Maurice JOANNES. The city of Dax paid homage to him by baptizing with his name the municipal stadium where his statue still stands guard today.

Above: One notices on this alignment of the SPAD XIII of escadrille SPA 77 the airplane of Boyau is the only one to be equipped with rockets the Prior. Paul Delestre Collection

Above: Maurice Boyau (bare head, 3rd on the right) discusses with his squadron comrades before leaving on a mission. The 6th on the right is MdL Jean Sardier (15 victories). In the center of the photo, the officer in the foreground with the look in the wave is Captain Pierre Mouronval, 2nd captain of the SPA 77, who won two victories before being killed in action on May 30, 1918. Paul Delestre Collection

Above & Facing Page, Below: Boyau prepares to leave on mission (g) and poses at the controls of his SPAD XIII (h). Paul Delestre Collection

Facing Page, Above: From left to right: X, Cpt Joseph Batlle (4 victories), MdL Jean Sardier (15 wins), X, Cpt Pierre Mouronval (from back, 2 victories, killed May 30, 1918), Sgt Henri Guillaume (1 victory), S / Lt Maurice Boyau (35 victories), X. Paul Delestre Collection

Above: Maurice Boyau, 5th left, captain of the French military rugby team. Paul Delestre Collection

Below: Pilots of SPA 77, the sportsmen squadron, pose behind a SPAD XIII. From left to right: X, Sgt Jean-Marie de Laubié (killed July 23, 1918, 1 victory in another squadron), Adjutant Francis Guerrier (4 victories), Lieutenant Jean Sardier (15 victories including 12 with SPA 77), S / Lt Maurice Boyau (killed September 16, 1918, 35 victories), Cpt Henry Decoin, squadron leader and French swimming champion (4 victories), Sgt André Géhin (killed May 29, 1918, 3 victories) and Sgt Richard Méwius (killed July 22, 1918, 1 victory). The photo was taken between May 14, 1918 and May 29, 1918. Paul Delestre Collection

Above: One of the latest portraits of the ace, made in Biarritz during his last leave where he poses with all his decorations: cross of officer of the legion of honor, military medal, cross of war with 16 fins and a star, not to mention two foreign decorations. Paul Delestre Collection

Above: Photo of Boyau's claimed winner, Leutnant Georg Von Hantelmann from Jasta 15, posing here in front of his Fokker D VII. It is highly likely, given his 500-meter firing distance observed by French infantrymen, that Boyau was shot down by a machine gun on the ground. Greg Van Vyngarden

Above: German Drachen; they were always strongly defended by batteries of machine guns. Greg Van Vyngarden

Maurice Boyau Victory List								
N°	Date	Unit	Plane piloted	Plane claimed	Place	Notes	German Casualties	Confirmation
1	3/16/17	N 77	Nieuport	Aviatik C	S. Thiaucourt		FlAbt 279 (A) - Uffz Karl Kolrop & Ltn Feodor Kellner killed in Remen-auville	Citation OA - JORF 23-07-1917 p.5699
2	6/3/17	N 77	Nieuport	Drachen	Geline	With MdL Sardier		Citation OA - JORF 28-09-1917 p.7659-7660
3	6/5/17	N 77	Nieuport	Drachen	Moussey			Citation OA - JORF 23-08-1917 p.6720
4	6/24/17	N 77	Nieuport	Drachen	Nancy-Goin	With S/ Lt Boudoux d'Hautefeuille and Sgt Boillot		Citation OA - JORF 30-10-1917 p.8689
5	6/24/17	N 77	Nieuport	Fighter	Nancy-Goin			Citation OA - JORF 30-10-1917 p.8689
6	7/13/17	N 77	Nieuport	LVG C	Coin sur Celles	With S/ Lt Boudoux d'Hautefeuille (2).	FlAbt A 276 - Uffz Rühl (unhurt) & Ltn Seifert (wounded)	Citation OA - JORF 02-12-1917 p.9729 "5th victory" (WRONG, 6th)
7	8/26/17	SPA 77	Nieuport	Drachen	Juvelize-Bourdonnaye			Citation OA 30-10-1917 - JORF 06-01-1918 p.271 "6th victory" (WRONG, 7th)
8	9/16/17	SPA 77	Nieuport	Drachen	Haie-Vauthier	With S/Lt Rebourg		Citation OA - JORF 29-12-1917 p.10684 "7th victory" (WRONG, 8th)
9	9/22/17	SPA 77	Nieuport	Drachen	Cirey-Bertrambois	With S/Lt Rebourg		Citation OA - JORF 02-01-1918 p.107 "8th victory" (WRONG, 9th)
10	9/23/17	SPA 77	Nieuport	Two-seater	Coincourt			Citation OA - JORF 02-01-1918 p.107 "9th victory" (WRONG, 10th)
11	10/1/17	SPA 77	SPAD VII	Two-seater	N. Champenoux	With S/ Lt Boudoux d'Hautefeuille and Cpt Tourangin (N 89).	FlAbt 12 - Ltn Hans Frowein & Ltn Oskar Sigg killed in Champ-enoux	Citation OA - JORF 31-12-1917 p.10771 "6 Drachen and 4 planes" (WRONG, 11th)

Maurice Boyau Victory List								
N°	Date	Unit	Plane piloted	Plane claimed	Place	Notes	German Casualties	Confirmation
12	1/3/18	SPA 77	SPAD	Drachen	Beney			Citation OA - JORF 04-04-1918 p.2936 "12th victory"
13	2/20/18	SPA 77	SPAD	Drachen	Vaxy	With S/Lt Sardier		Citation OA - JORF 04-04-1918 p.2936 "13th victory"
P1	5/28/18	SPA 77	SPAD XII	Albatros D	SE Fismes			
14	5/29/18	SPA 77	SPAD XII-Canon	Drachen	Bois de Dole			Citation OA - JORF 24-07-1918 p.6381 "14th victory"
15	5/29/18	SPA 77	SPAD XII-Canon	Fokker D VII	Ville-en-Tardenois	With S/Lt Sardier		Citation OA - JORF 24-07-1918 p.6381 "15th victory"
P2	5/29/18	SPA 77	SPAD	Albatros D				
16	6/1/18	SPA 77	SPAD	Pfalz	Epieds			Citation OA - JORF 20-08-1918 p.7339 + JORF 24-09-1918 p.8327 (1st) "16th victory"
17	6/4/18	SPA 77	SPAD	Drachen		With S/Lt Sardier		Citation OA - JORF 20-08-1918 p.7339 "17th victory"
18	6/4/18	SPA 77	SPAD	Drachen		With S/Lt Sardier		Citation OA - JORF 20-08-1918 p.7339 "18th victory"
19	6/27/18	SPA 77	SPAD	Drachen				Citation OA - JORF 24-09-1918 p.8327 (1st) "19th victory"
20	7/1/18	SPA 77	SPAD	Drachen		With S/lt Haegelen (SPA 100)		Citation OA - JORF 24-09-1918 p.8327 (3rd) "20th victory"
21	7/5/18	SPA 77	SPAD XIII	Drachen	Ville-en-Tardenois			Citation OA - JORF 24-09-1918 p.8327 (2nd) "21th victory"
22	7/5/18	SPA 77	SPAD XIII	Fighter	Ville-en-Tardenois			Citation OA - JORF 24-09-1918 p.8327 (2nd) "22th victory"
P3	7/5/18	SPA 77	SPAD	Enemy plane				
23	7/15/18	SPA 77	SPAD	Fokker D VII	S.E. Dormans			Division Aérienne logbook ("23th victory") + Citation OA - JORF 24-09-1918 p.8327 (3rd) "23th victory"

Maurice Boyau Victory List								
N°	Date	Unit	Plane piloted	Plane claimed	Place	Notes	German Casualties	Confirmation
24	7/17/18	SPA 77	SPAD	Enemy plane	N.E. Nesle-le-Repons			Division Aérienne logbook ("24th victory")
25	7/21/18	SPA 77	SPAD	Enemy plane	S. Soissons		Jasta 26 - OfStv Otto Esswein (12), killed in Hartennes, 15 km S Soissons	Division Aérienne logbook ("27th victory") - Late Confirmation
26	7/22/18	SPA 77	SPAD	Drachen		With Adj Guerrier		Division Aérienne logbook ("25th victory") (WRONG, 26th)
27	7/22/18	SPA 77	SPAD	Drachen		With Adj Guerrier		Division Aérienne logbook ("26th victory") (WRONG, 27th) - Citation OA (OLH) - JORF 23-10-1918 p.9193 "27 victories, 16 Drachen and 11 airplanes"
28	7/22/18	SPA 77	SPAD	Enemy plane	Fresnes			Division Aérienne logbook ("28th victory")
29	8/8/18	SPA 77	SPAD	Drachen	Cury			Division Aérienne logbook ("29th victory")
30	8/8/18	SPA 77	SPAD	Rumpler C	Piennes	With MdL Lentz-Mitchell		Division Aérienne logbook
31	8/8/18	SPA 77	SPAD	Two-seater	Cury / Ressons-sur-Matz			Division Aérienne logbook ("31th victory")
32	9/14/18	SPA 77	SPAD	Drachen	Etraye	With Cprl Corsi and S/Lt Haegelen (SPA 100)		Division Aérienne logbook
33	9/15/18	SPA 77	SPAD	Drachen	La Haie des Allemands	With S/Lt Barbaza, Lt Decoin and Adj Strohl		
34	9/15/18	SPA 77	SPAD	Drachen	Foulgrey	With S/Lt Barbaza, Lt Decoin and Adj Strohl		
35	9/16/18	SPA 77	SPAD XIII n°10666	Drachen	Harville	With Asp Cessieux		Division Aérienne logbook - Individual file: "35 official victories" (Note of GC 17 commanding officer)

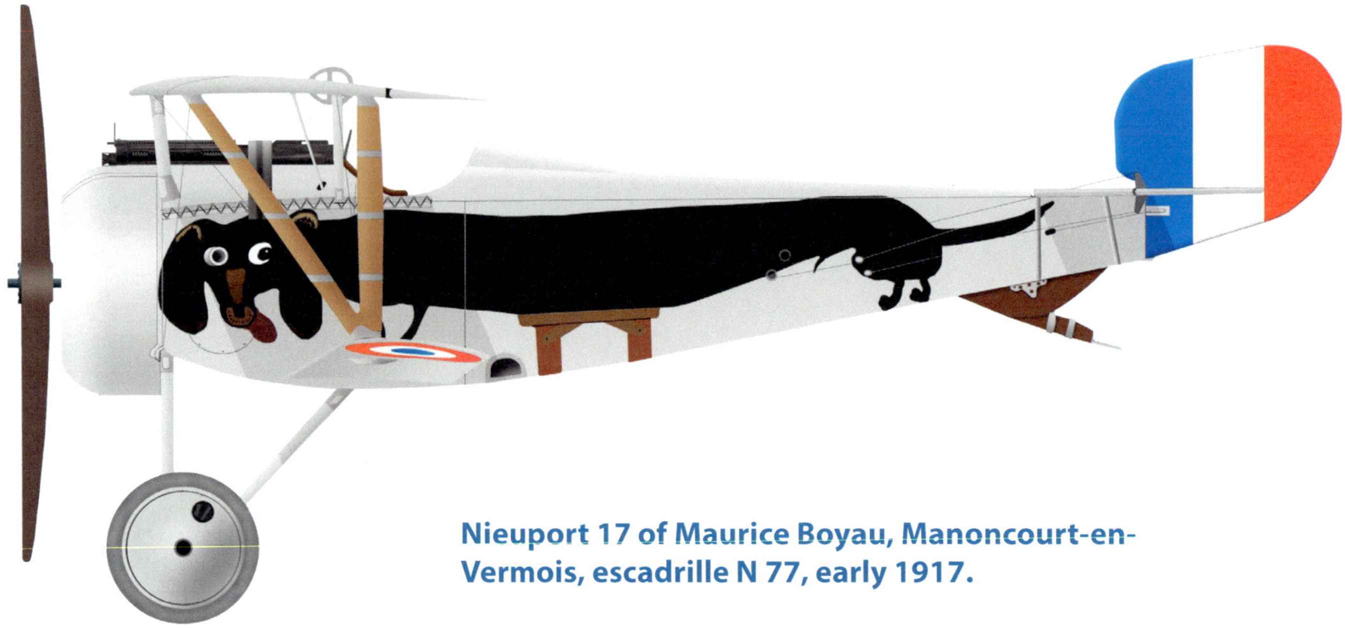

Nieuport 17 of Maurice Boyau, Manoncourt-en-Vermois, escadrille N 77, early 1917.

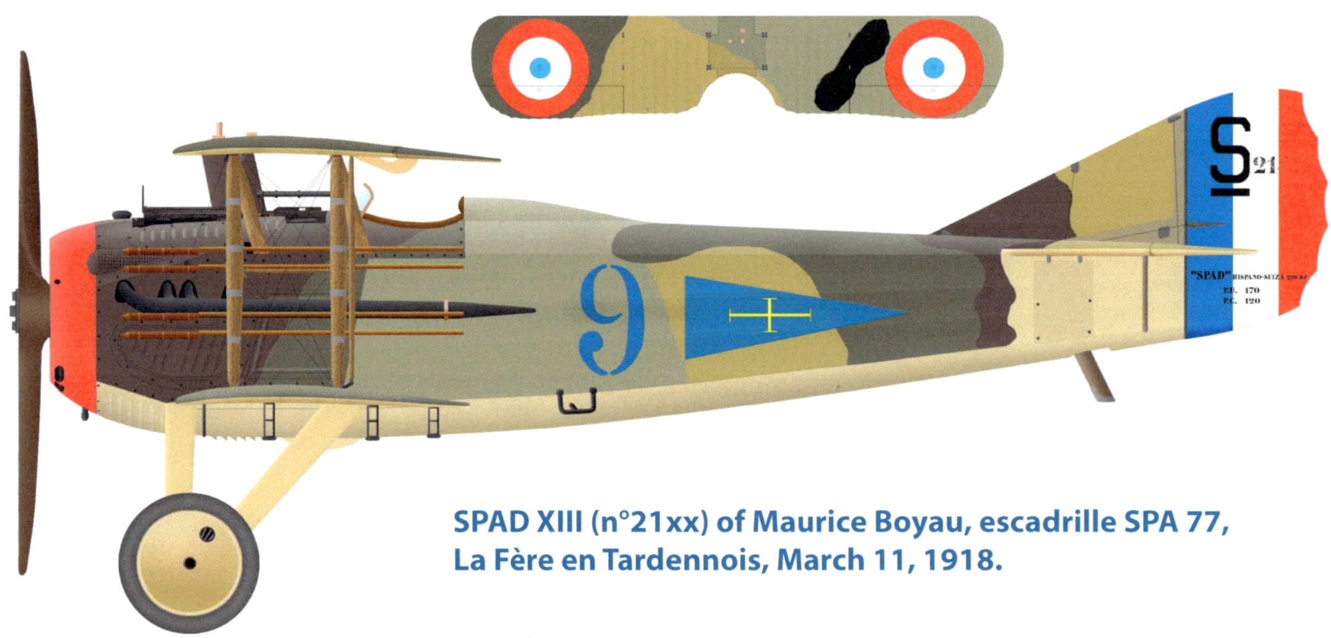

SPAD XIII (n°21xx) of Maurice Boyau, escadrille SPA 77, La Fère en Tardennois, March 11, 1918.

Above: Sergent André Boillot in front of the model of Maurice Boyau's Nieuport 17 personnal decoration ?

Above: This plane has been presented as Maurice Bouyau's personnal plane. There is no clue to illustrate this fact ; all N 77 Nieuports in 1917 had colorful personnal decorations.

Above & Below: Captured German planes photographed in Paris, place de la Concorde, in October 1918. Photo credit Auguste Léon - Public domain.

Made in the USA
Monee, IL
12 September 2026

62974515R00136